SYNERGETIC THEORY OF LIFE

SYNERGETIC THEORY OF LIFE

Nature

Algorithm

Self–Insight

Conscious Evolution

ORAZ TURKMEN

Matador
9 Priory Business Park,
Wistow Road, Kibworth Beauchamp,
Leicestershire. LE8 0RX
Tel: (+44) 116 279 2299
Fax: (+44) 116 279 2277
Email: books@troubador.co.uk
Web: www.troubador.co.uk/matador

ISBN 978 1780881 096

British Library Cataloguing in Publication Data.
A catalogue record for this book is available from the British Library.

Typeset in 11pt Bembo by Troubador Publishing Ltd, Leicester, UK
Printed by CPI Group (UK) Ltd, Croydon, CR0 4YY

Matador is an imprint of Troubador Publishing Ltd

In bright memory of mother and father

Contents

Foreword

The first time we read this book, it was filled with many difficult concepts and ideas that just seemed out of our reach. It was tempting just to put it aside and blame its depth and extent. However considering the amount of work that went into this book, we had to do our best to understand it. Indeed the author, our grandfather, spent not months or years but decades developing the Synergetic Theory of Life.

Being as complex as it is, we think that this book is spot on when talking about the problems of the current world. We are not talking about naming precise issues but about the root of mankind's current problems. It may sound familiar with many other books trying to pinpoint the issues in our world and their solution. However we believe that this book is on a different level. We can bravely say this because we knew the author very well. He was not someone trying to get his name out there but was a dedicated scientist with decades of scientific work under his belt. This is one of the reasons why you should give this book a try.

In our opinion, this book describes well where we, as mankind, are and where we are headed to. In the modern fast paced world no one stops and thinks about where we are headed to as a civilization and most importantly as a species (we are indeed a part of nature). If we stop and look at the current state of the world we will find that it is on a path of destruction – natural phenomena like hurricanes, droughts, floods; environmental misbalance like the climate change, deforestation; resource famine; economical disasters and the increasing gap between poverty and wealth, severe recessions; social, political and of course religious conflicts that have caused numerous wars. All of these are not just everyday happenings but they have an extreme influence on the Earth, nature, mankind and Life on Earth as a whole. It is important to stop and think about where those happenings can lead us to. Therefore the author, understanding the grave significance of the above, has created the Synergetic Theory of Life. He shows the evolution path of Man and Life from a new perspective and proposes the new path of future development that mankind should take in order to avoid self-destruction, which is inevitable if we carry on as we do currently. Mankind, whatever way you see it, whether as the greatest creation of nature or simply as people you see every day, has to realize its own destructiveness and change onto the new path – the path of change of its current ways. Changing like that is not a simple task and hence this is not a simple book.

After we read this book a few times we began to view the world differently and thought more about the future rather than the present. We hope this book will be useful to you and by reading it you too will learn something new and change something in yourself.

Kemal Atayev
Student of University of Birmingham

Ovez Atayev
Student of St Edward's School, Oxford

September 2011

Introduction

Today's destructiveness is a characteristic feature of the history of humanity. It creates a real threat to the existence of Life on the Earth. Man's habitat becomes less comfortable and sometimes even aggressive: every year six million hectares of rich soils turn into infertile deserts; eleven million hectares of forest are extinguished; some whole biologic populations and species vanish from the face of the Earth.

Great problems have also arisen in agriculture: droughts, floods and various diseases of farm crops have become usual and do great harm to agricultural production. Low quality of food and its physiological misbalance lower Man's quality of Life.

The Earth's atmosphere and hydrosphere become more and more polluted. The greenhouse effect, a thinning of ozone layer of the Earth's atmosphere, global warming, the melting of Arctic and Antarctic ice and finally, the climate change becomes the reality of life.

An ever-growing number of catastrophes are observed in the sphere of Man's *being*: food, power and raw-material famine; ever-increasing differentiation of social level of peoples life; the growing gap between poverty and wealth; overpopulation of the globe, illiteracy; danger of distribution of nuclear, chemical, biological and other kinds of weapons of mass destruction. Thus Man's discomfort becomes more explicit and his dissatisfaction with reality more pressing which awakens his desire for a cardinal change to this reality. With this in mind, various forms of reformation of available and creation of new, rather immature social orders are suggested which do not originate from the Man's nature, but from the problems of his *being*. The communist, chauvinist, positivist and other analogous forms of ideologism, the black-and-white representation of reality (and radicalism being characteristic of them), all take a special place there.

The latest history has shown rather convincingly what the first two of the above forms of ideologism represent. As to the positivistic ideology, which is ruthlessly befuddling people, it turns them into seekers of materialism people, and makes them slaves of technology and consumption.

It is time for Man to understand that, under the supremacy of positivism and pragmatism, both ideology and politics as the forms of society government have exhausted themselves and are becoming cynical and potentially dangerous for a society. Politics, as the art of government control, acts here as a balance of interests without allowance for mentality, moral principles, level of intelligence development, etc., which humanity has not completely realized, and people continue to see the improvement of their life quality in using newer and newer imperatives of *being*.

The imperative of cultural revolution, ecological imperative, technological imperative, imperative of sovereignty, imperative of economic growth, democratization imperative, globalization imperative and imperative of law – a far from a complete list of imperatives of *being*, defined only in the last few decades. In principle a list of analogous imperatives can include hundreds and thousands of them, but in any case it cannot be a resolution of the problem, since Man's *being* is of existential character and does not yield to logical formalization in a form of scientific categories, notions, etc. Besides, as the number of the used imperatives grows, the number of their mutually exclusive objectives will grow as well, and will finally make it all pointless.

It should be noted that among new imperatives of *being*, new projects of social system are proposed: post industrial society, open society, civil society, informational society, society of sustainable development, etc. This list may be continued, but it is obviously a dead end, since any attempt to improve the Life quality turns here into an endless string of complex and mutually exclusive problems of Man's *being*.

In the above context it is evident that the understanding of Man's nature is always being simplified. The existent ideas and conceptions are most variant (Man is a divine creation; Man is Homo sapiens; Man is the complex of social relations; Man is an animal; Man is a robot; the econom-Man; inform-Man, etc.), but they reflect only separate aspects of the problems of Man's *being*. That of course is a symptom of the crisis situation which has arisen, since there is no full or complete understanding of a Man.

At the same time the significant problem of the future of humanity becomes more and more acute and topical. A reasonably broad mass of people get an acute feeling of an approaching Apocalypse: ecological crisis; problem of climate warming and its ever increasing anomalies (hurricanes, floods, droughts, melting of the Arctic and Antarctic ices, excessive cold and heat); food, power and raw-material famine; technological catastrophes; problem of the Earth's overpopulation; problem of possibility of psychological perception of a politicized classical myth of colliding civilizations (cultures) by broad masses of people; degradation of traditions, moral and ethic values and other analogous problems become a bitter reality today and create a principally new kind of fear in people – fear of the future.

But the most important and dangerous thing is that the contemporary Man's psyche suffers from the increasing strains and becomes predisposed to percept the idea of the destructiveness of Life's existence on Earth. This confirms that contemporary Man, in his evolutionary development, has reached a critical state and his further existence really seems problematic. It is possible (evident) that the way to his salvation is in revealing the secret of the nature of Life and the nature of a Man himself. However, these problems become extremely complicated since there is almost no time for their solution.

The epoch of the non-conscious history of humanity seems to be over, and it is now necessary to lay the foundation of a new epoch, that is the epoch of the conscious history where the nature of Life and Man will not be a secret of the universe and this will open fundamentally new horizons of development common to all mankind.

The scientific work brought to readers' attention is devoted to these extremely complicated and fortune-bearing problems. One should take notice of some specific peculiarities of the work's design.

The problem of Life and the problem of Man are so extensive, that they are far beyond the limits of any field of science. Neither natural, nor humanitarian and technical fields of science can embrace them on their own. That is why the categories and notions used in these scientific spheres are, as a rule, fragmentary, ambiguous and reflect only peculiarities of a paradigm and methodology of these particular fields of knowledge, which does not provide clarification, but complicates the problem and leads it to a dead end. To remove these principal difficulties the author uses neologisms, i.e., new terms and notions, as well as new meanings of the existing terms and notions and algorithmic approach to analysis, which allowed him to create and state compactly the Synergetic Theory of Life (STL) and to develop a scientific forecast for the nearest future. So, when studying this work one should use the author's terminological dictionary placed at the end of the book as a supplement. The reference to this dictionary is indicated by an asterisk.

Also, because of the versatility and extensiveness of the issue under study and difficulties in following accepted practices, the author has abandoned the traditional method of citing specific scientific literature, and proceeded instead from generally accepted scientific propositions, facts and theoretical knowledge.

CHAPTER 1

Nature and Algorithm of Life Evolution

For Man the cognition of Life's nature always was and still remains the foundation-stone of his existence. The myths and religious notions predominated in this sphere up to the 18th century. The East ethic-religious teachings: Buddhism, Daoism, Confucianism, Sintoism, Jainism, and others, which preach the idea of integrity of Man's existence in the world, here stand by themselves. Meanwhile the problem of Man's nature is substituted, by mistake, by the particular problems of Man's *being*, and Man's social development is subject to stagnation. And only in the last 300 years has the problem of Life's nature and Man's nature become a subject of scientific research. A list of the most wide-spread ideas and conceptions concerning a Man is given in the Table as an example. As is evident from the Table, there are a lot of various approaches to this problem's solution: from traditional-religious and religious-ethic notions to exact technical models.

The modern science and scientific-and-technological progress, on the one hand, have considerably expanded the notion of Man, simplifying its understanding to the level of a robot, and, on the other hand, they have not brought the necessary clarity to the understanding of the nature of Life in general, and Man's nature, in particular. And the main thing is that these fundamental problems were often reduced by mistake to the more particular problems of Man's *being* and thus a dead-lock in their understanding was created. This deepest delusion of the contemporary Man, still dominates in his world outlook. Meanwhile, these problems are of the primary and fundamental importance for understanding of Man's evolutional development, since these specific problems, rather than the problems of *being* should be the basic ones in understanding of realities.

TABLE

Approximate list of the existing notions, ideas, conceptions and models of a Man

Sources	Content, essence	Note
1.Mythologic, religious	Man – the greatest creation of God	Creationism, Judaism, Christianity Islam
2.Philosophic, ethic	Man is a being and a consciousness	Spiritualism, Carthusianity, Dialectic materialism
	Man is a clever being or a man is a rational animal, Homo sapiens	Anaxagoras, Plato, Aristotle
	Man is a social animal	E. Kant
	Man's nature is evil personified	Buddhism
	Man's nature is goodness personified	Humanism, Humanistic Psychology
	Man – is a creature determined by attractions. Positivist	Positivism, Pragmatism
	Man as "Life disease", or a Man is the evolutional deadlock.	Panromanticism
	Superman	F. Nietzsche
	Man is a spiritual being, free of vital dependence and open to the world	M. Scheller
	The world, including a Man, is the opposition of two "eternal principles" – the Good and the Evil	Zoroastrism
3.Natural-Scientific	Man is a product of bioevolution of animals	Physiology, psychology
	Zoomorphic models of a Man (ratmorphism, dogmorphism, monkeymorphism, etc.);	E. Cassirer C.G. Young, L. Bertalanffy
	Man is a symbolic animal; A Man is a symbolic creature	

4.Scientific-Technical	Man is a machine	D. Lametra
	Man is a robot	K. Capek
	Man is a component (element) of technology; artificial intellect	Cybernetics
	System-symbolic conception of a Man	L. Bertalanffy
	Informational model of a Man (inform-man); synergetic-informational model of a Man	Molecular biology, information technology
5.Economic, social	Man – a set social relations	Marxism-Leninism
	Man – a basis of economy and economic relations; econom-man	A. Smith, Marxism
	One-Dimensional man	H. Marcuse
6.Psychophysio-logical	Bioman, determinated by unconscious	Psychoanalysis, Freudism
	Bioman whose behaviour is determined by the pattern "stimuli-response"	Behaviourism, D. Watson, B. Skinner
	Selfactualizing Man	A.Maslow, K.Rogers
	Man – a social animal	C.G. Young
	Man's intellect is not a product of evolution, but an extraterrestrial phenomenon	N.P. Bekhtereva

One should not forget that Man's *being* is based on the canons of Life evolution, rather than vice versa. Besides, in the course of Life evolution the form of *being* is constantly changing, and the *being* is determined by the used way of adaptation. In this connection, all the available diversity of ideas and conceptions concerning a Man cannot be considered as a certain secondary question, since it is a bright manifestation of the arisen evolutional crisis and destructiveness of Life existence. The way out of this apocalyptic situation is a more profound understanding of Life's nature, Man's nature and the canons of their evolutional development.

The first and subsequent scientific theory of Life – the theory of biological evolution was proposed in 1809 by French naturalist and philosopher Jean Battist Lamarck. He proceeded from a supposition that all living beings have the long-term history of continuous change due to the action of the laws of nature. A mechanism of such variability was seen by him in the inheritance of acquired features. But unreality of this supposition had been later proved by science.

In the context of modern science one can notice that as genetic inheritance of acquired features is truly impossible, evolution has found another absolutely new channel of information transmission which can help a Man to acquire the experience and knowledge gained by previous generations of people. This is not biological evolution anymore but evolution of a new type, that is, the intellectual evolution of Life.

J.B. Lamarck's study had been further enriched and developed by Charles Darwin. The latter gave attention not only to the time factor of evolution of living beings, but also to their diversity, i.e., to multiplicity of populations and species, and proposed a new mechanism of biological evolution – the natural selection. The 's theory was published in its full volume on as the book *On the Origin of Species*. Undoubtedly, Darwinism had extended the horizons of cognition of the world of living nature by a Man, and that is why the 19th century was called the century of C. Darwin.

The theory of evolution by C. Darwin, being a foundation of the current biologic science, further on was continuously specified and perfected that resulted in the synthetic theory of evolution. It includes achievements of not only classical biology but also chromosome theory of inheritance, population genetics, and molecular biology. And this new level of theoretical generalization also completely denies a possibility of genetic inheritance of acquired features and shows that evolutionary phenomena occur at the level of populations and biological species rather than on the individual level of separate individuals as stated by Darwinism.

If Darwin's theory of evolution proceeds from the fact that natural populations consist of individuals of more or less the same genetic type (however accepting the fact of the existence of several different mutations among them), then the synthetic theory of evolution is based on the fact that these natural populations possess a great reserve of genetic variability, and evolution uses finer mechanisms than was previously thought.

The laboratory experiments have indeed confirmed that the higher the level of genetic variability in a given population, the more rapid is its evolution. The synthetic theory dominates nowadays when considering all the aspects of biological evolution, but being a fundamental achievement of modern science, it still remains a fragmentary and un-completed one. This, first of all, concerns the initial and the highest stages of evolution. In so doing, we assume that the synthetic theory develops so far in isolation from the problem of appearance of first living organisms which were the ancestors of every living thing on the Earth.

The situation is especially complicated with transition to the highest stage of Life – a Man. The questions: is Man evolving, how had the human intellect arisen, what does mankind expect in the nearest future and many others as essential – remain unanswered.

There is no doubt that the origin of Life on the Earth was preceded by formation of hydrosphere and primary atmosphere which consisted of methane, ammonia, water vapor, and hydrogen. Then, as a result of the effect of electric discharges, corpuscular and ultraviolet Sun radiation and, maybe some other physico-chemical factors on the Earth's atmosphere, there arose a complicated set of abiotic organic chemical compounds.

The further accumulation of the latter in a water environment was the origin of chemical evolution which led to the formation of noncomplex low-molecular (monomer) and then high-molecular (polymeric) compounds of protein substances and nucleic acids.

The next important step was the formation in a water environment (in

Fig.1 Life evolution algorithm

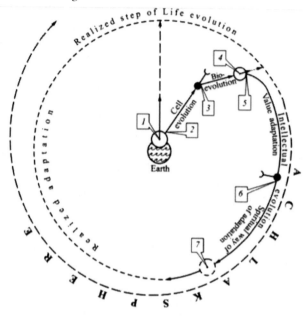

	Symbols	1 – diversity of chemical composition and physico-chemical conditions; primary abiotic organic substances and origin on their basis of low-molecular compounds (monomers); enzymes, polymeric compounds (proteins and nucleic acids), i.e. polymeric broth.
	bifurcation of the environment – appearance of principally new method of Life existence	2 – life, as synergism of nucleic acids and proteins; adaptability; bifurcation of environment, i.e. bifurcation of physico-chemical environment; environment bifurcation I;
		3 – adaptational bifurcation I; populational adaptability;
	bifurcation using the method of adaptation	4 – unique biodiversity;
		5 – perception; value adaptability; medium bifurcation II. i.e. bifurcation of biologic enironment;
	evolutional deadend	6 – adaptational bifurcation II; spiritual way of adaptation; apperception;
		7 – conscious intellectual bifurcation; conscious adaptation

shallow water, on the shore plots, on water surface, etc.) of specific plots where with the growth in concentration and complication of composition of organic substances the environment becomes unstable and its bifurcation takes place due to Synergism (cooperation) of nucleic acids and proteins. Principally new synergetic (self-organizing) structure formed as a result of this process was the act of Life's origination on the Earth. Self-reproduction and adaptability to varying conditions of the environment are its qualitatively new and unique attributes. It is these that became the basis of Life evolution in the future. Coming from this notion the author has developed the Synergetic Theory of Life (STLifeife), which is shown in algorithmic form in Fig. 1.

According to the synergetic theory the era of chemical evolution, which had led to Life's origination, probably began soon after the Earth's formation – 4.6 billion years ago. Approximately one billion years later Life* originated, as a product of chemical evolution, in a form of self-organizing structure, i.e. a protocell capable of adapting itself to varying conditions of the environment. Then the cell evolution of Life continued for three billion years. During this time the protocells evolved into prokaryotes, i.e. anaerobic anoxybionts (microorganisms, that use anaerobic, i.e. oxygen-free photosynthesis).

The asexual reproduction was characteristic of prokaryotes, which already had membranes which isolated them from the environment. This has qualitatively changed metabolic processes, and they became the basic mechanism of the further evolution of prokaryotes.

That is why the cell evolution was in essence not only biochemical but also a metabolic evolution of Life. It was completed by the origin of eukaryotes – cells with a nucleus which consists of chromosomal elements. Herewith the anaerobic photosynthesis was a primary source of cell energy, i.e., fermentation of abiotic organic substances, which led to the discovery of secondary atmosphere by evolution. The secondary atmosphere consisted of nitrogen, oxygen, water vapor, carbon dioxide and an ozone layer protecting living organisms from destructive ultraviolet solar radiation; the process of oxidation, i.e., aerobic photosynthesis became the basic source of energy for eukaryotes.

Eukaryotes, being the aerobic oxybionts, have radically changed the character of the evolutional process. In particular, the sexual cells were discovered in the process of their evolution. And their appearance entailed the inclusion of absolutely new mechanism in the evolutional process – sexual reproduction of organisms, evolution of which revealed the phenomenon of reproductive isolation and, respectively, a qualitatively new step of Life adaptation, that is, morpho-physiologic or population adaptability★.

Sexual reproduction instead of metabolic mechanism of evolution puts into action a new unique mechanism – morpho-physiological mechanism of Life evolution. Thus sexual reproduction, based on synergism of male and female sexual cells, is a triumph of cell evolution of Life and, as to its value, is equivalent to the discovery by evolution of archetypical memory of a Man.

The efficiency of populational adaptability★ proved to be incomparably higher than primary Life adaptability, thus, a new type of bifurcation took place in the process of evolution; it may be called adaptational bifurcation I. This evolutionary metamorphosis happened approximately six hundred million years ago and initiated the beginning of classical biological evolution. In geologic time scale it corresponds to Phanerozoic time, i.e., the beginning of the Cambrian period.

A new mechanism of adaptation – populational adaptability – radically changed the character of evolutional process: multicell organisms appeared; there occurred an "explosion" in diversity of living organisms. The evolution rate has become incomparably higher: the Life area has been spread not only to the hydrosphere, but also to the land, including such extreme territories as mountains, deserts, glaciers, embracing the whole Earth globe. And above all, as was mentioned above, an incomparable increase of the rate of evolution happened, its peak being in the appearance of hominids who had a relatively developed reflexion and moved on two limbs.

The unique biodiversity, which appeared in the process of biological evolution, including hominids, possessed great potential for further evolutional development of Life which was then realized approximately four million years ago by bifurcation of the biological environment (environment bifurcation II). This evolutional metamorphosis occurred at the level of hominids by synergism of receptive senses and the discovery of a principally new way of reality perception through sensual images, i.e., the perception★ phenomenon. *Homo perceptions* has become the owner and carrier of this quality; due to this asset *Homo perceptions* has acquired unique new charecteristics, that is: types of senses peculiar to Man only. Anthropithecus were anthropologic equivalents of *Homo perceptions*.

In terms of morpho-physiology the perceptional (unconscious) perception is none other than the appearance, by the evolutional way, of new specific structures of brain cortex, implanted in its primary archaic layer and responsible for instincts.

Thus *Homo perceptions*, possessing the perceptional perception and primary human senses, that is relic archetypes, has in its evolutionary development risen to qualitatively new level of intellect, and thus a new stage of evolution was opened which can be called the intellectual evolution of Life. In other words, owing to bifurcation of the biological environment (environment bifurcation II) the biological step of Life evolution had developed into intellectual one, i.e. the evolution of a   When summarizing the above mentioned, one can distinguish the following absolutely important propositions.

1. Life has a synergetic nature and its self-reproduction and adaptability are the primary fundamental natural attributes.
2. Life evolves following the adaptation-bifurcational mechanism rather than the mechanism of natural selection. Life evolution is both continuous and discrete. Meanwhile the bifurcation of the environment generates the phyletic type of evolution, and the adaptational bifurcation – species-

forming type of evolution. The rate of the first type is much (hundreds of times) slower than the rate of the second type of evolution.

3. Biological evolution occurs at the genetic level through the origin of new biological species. The cellular and biological evolutions, being the continuation of the chemical one make up the first step of Life evolution. The essence of Life evolution is reflected in the following diagram:

synergism → origin of a new self-organizing structure → bifurcation of the environment.

4. Continuation of the biological step of evolution is the intellectual step of Life evolution, i.e. Man's evolution based on archetypical (otherwise, psycho-physiological) mechanism of intellectual development.

Consider in more detail Man's evolution. Its algorithm is presented on Fig. 2. As was noticed above, it begins from the environment bifurcation II, i.e., bifurcation of the biological environment which resulted in the origin of *Homo perceptions*. This event took place approximately three million years ago. His dominant functions were perception (the perception of reality through sensual images, i.e., human senses) and value adaptation★. The volume of *Homo perceptions'* brain was about 750 cm³.

If metabolic and genetic mechanisms were the determining ones at the first step of Life evolution, then Man's evolution is based on the archetypical, or psycho-physiological mechanism. The fact is that, thanks to the perceptive ability of *Homo perceptions*, the relic archetypes★, being sensual images of his instincts were opened in the course of his evolution. And these relic archetypes further resulted in the origin of his archetypical memory. In connection with this *Homo perceptions* rose to a new hierarchic level of intellectual evolution, i.e., he became a new intellectual species – *Homo habilis* capable to sense-in (understand). The volume of his brain increased a bit – to approximately 1100 cm^3. He did not eat raw meat anymore, but that cooked on fire. This evolutional metamorphosis occurred about a million and a half years ago. In anthropologic aspect Sinanthropus correspond to this stage of Man's evolution. They, like pithecanthropus, did not yet speak, and social condition of their existence was gregarious.

BIOLOGICAL EVOLUTION

HOMINIDES: ramopithecus, ..., australopithecus

Dominant attributes:

1) movement on two limbs;

2) relatively developed reflexion;

3) t ~ 4m years; V_{BV} ~ 450 cm^3

Biological evolution through synergism of receptive senses opened *anthropithecus,* a new form of Life existence. The relic archetypes, being sensual images of biological instincts, were his unique quality. It is due to this evolutional metamorphosis anthropithecus became able to perceive that, in turn, led to bifurcation of the biological environment (environment bifurcation II). That is why to distinguish this specific feature of anthropithecus we shall further call him *Homo perceptions.* In terms of morpho–physiology, perception is the display of the phenomenon of integrity of human brain function, which opens a possibility of the world perception with the help of sensual images, i.e., types of senses inherent only in a Man, that is the human senses. And relic archetypes, as sensual images of instincts, have led to the discovery of "sensing-in" (understanding) phenomenon, and respectively, qualitatively new type of intellectual ability in *Homo perceptions.* As a result, the biologic evolution found its continuation in the intellectual evolution, i.e., Man's evolution, as the next step of Life's evolution.

HOMO PERCEPTIONS
(anthropithecus)
Dominant attributes:
1) perception, perception of the world through sensual images (or human senses);

2) relic archetypes; value adaptation;

3) t ~ 3m years; V_{BV} ~ 750 cm^3

4) anthropologic equivalent:

Australopithecus

Discovery of phenomenon of long-term memory by intellectual evolution

HOMO HABILIS (archeoanthropus, paleoanthropus)

Dominant attributes:

1) "sensing-in" (understanding);

2) primitive labor tools; use of fire;

3) t ~ 1.5m years; V_{BV} ~ 1100 cm^3;

4) anthropologic equivalent: from sinanthropus to Neanderthal, being a transitional species to a contemporary Man

Intellectual evolution opened the archetypical phenomenon of the Self (Man's understanding of integrity of his existence in the world) and independence of brain (thinking). Adaptational bifurcation II.

HOMO SAPIENS

(neoanthropus, totem-religious Man)

Dominant attributes:

1) spiritual world outlook; spiritual way of adaptation;

2) apperception; archaic thinking; rational (mental) cognition;

3) t ~ 40 000 years; V_{BV} ~ 1400 – 1600 cm^3;

4) anthropologic equivalent: Cro-Magnon Man

HOMO SAPIENS SAPINES (religious Man)

Dominant attributes:

1) static world; static thinking;

2) religions; written languages;

3) t ~ 10 000 years; V_{BV} ~ const;

4) anthropologic equivalent; contemporary Man with religious mentality

POSITIVIST

Dominant attributes:

1) spiritual crisis; positivistic world outlook;

2) mental disorder (loss of mind); negative adaptation; estrangement; crisis of intellectual evolution;

3) t ~ the present; V_{BV} – const;

4) anthropological equivalent: true Man standing at the top of intellectual evolution

Self-insight; conscious bifurcation of mankind.

SELF-SUFFICIENT MAN
(superanthropus; individual)
Dominant attributes:

1) scientific religion; synergetic world outlook; scientific intuition;

2) conscious evolution;

3) t ~ !; V_{BV} – const;

4) future Man (individual)

HOMO ABSOLUTUS (perfect individual) Absolute perfection and Akhlaqsphere are ideals of *being*

Fig. 2 Algorithm of Man's evolution (VBV – brain volume, t – approximate time)

A unique quality of *Homo habilis* was his ability to sense-in (understand). In the course of evolution this quality was constantly developing and reached the level of understanding by him of the integrity of his existence in the world. Meanwhile, thanks to synthesis of relic archetypes there arose the archetypical phenomenon of Self, symbolizing the understanding by Man of the integrity of his existence. Further, the archetypical phenomenon of Self, being fixed in the genetic memory, resulted in the origin of the contemporary Man, and, respectively, there occurred the adaptational bifurcation II. In terms of morpho-physiology the Self phenomenon in ontogenesis is realized in the most external layer of brain cortex of the contemporary Man in a form of specific structural elements. That is why the Self phenomenon is a common characteristic to all mankind.

So, the synergism of relic archetypes has resulted in the origin of the contemporary Man★ that, in turn, has led to bifurcation of the way of adaptation, i.e., there occurred the adaptational bifurcation II. This evolutional event took place about 40,000 years ago. Cro-Magnon Men were the anthropologic equivalents of a contemporary Man.

The contemporary Man★ has three phases of evolutional development: *Homo sapiens* (totem-religious Man), *Homo sapiens sapiens* (religious Man) and Positivist Man. The volume of brain of a contemporary Man in the phase of *Homo sapiens* was about 1400-1600 cm³, and it has not changed since then.

A unique quality of the contemporary Man is the phenomenon of thinking★ being a result of Man's brain independence★. Thanks to thinking, Man's perception became apperceptive, i.e., intelligent, and the main thing is that he possessed a qualitatively new spiritual way of adaptation★. In other words, spirituality has become the determining factor in the evolution of a contemporary Man; this spirituality is based on Self which symbolizes the understanding by Man of the integrity of his existence in the world. With the appearance of *Homo sapiens* the intellectual evolution acquires a cognitive character. Such fundamental qualities of the contemporary Man as speech, morals, mysticism, myth-creation, magic, art, writing, philosophy, rational cognition, etc., were opened by evolution during a relatively short time. And his *being* became social instead of a gregarious one. It should also be noted that the rational cognition develops by alternation of the spiritual leap and the intellectual development, i.e., the awakened spirituality serves a basis for the development of intellect. This process is always repeated, and intellectual evolution is of stepped character on the whole.

Thus, the intellectual evolution, based on the archetypical (or psychophysiological) mechanism, proved to be exceptionally efficient, since Man's development occurred over a relatively short time (about three million years), otherwise equal to hundreds of millions of years.

Currently Man's evolution is in deep crisis, since he destroys the archetypical phenomenon of Self and, respectively, loses his religious spirituality and as a result he is increasingly transformed into the phase of positivist Man (in the language of religion – into the devil, melgun), who uses negative adaptation instead of the spiritual way of adaptation; the negative adaptation is directed towards Man's supremacy over nature. Therefore the following are peculiar to him: "locking-in" on himself, estrangement, mercantile attitude to the outside world and "civilized" (technological) barbarism. It is these factors that cause all the anomalies and the destructiveness of the existence of the contemporary Man.

One can say surely today, that it is Man himself who has become a deadly disease of Life on the Earth. Having lost his religious spirituality, having exhausted his evolutional potential, he has found himself in the dead-end of Life evolution. And even though Man is not still conscious of that, he feels it intuitively. What is more, to get out of this critical situation, basing on empirical sociology and his own intuition, he tries to follow the way of reformation of his *being*. But this is the way to nowhere, since the reason for the critical situation which had arisen is not in *being*, but in the evolutional crisis of Man himself. Man's behaviour becomes more and more feverish and creates the feeling of future shock. To preserve Life on the Earth, Man has to become absolutely different. A conscious intellectual bifurcation and the appearance of a new species – self-sufficient Man★ now becomes Man's evolutional destiny.

CHAPTER 2

Self-Insight and Conscious Evolution

Man's intellect origin is the evolutional event equal to the origin of Life itself. Thus, let's try to look at this event in more detail. Fig. 3 shows the algorithm of Man's intellect evolution created by the author. Big circles mean Man's intellectual species and intellectual populations, circles on the left side – the essence of the evolutional phase, circles on the right side – human emotions and senses.

A fundamental principle of human intellect was a phenomenon of perception*, which arose through evolution as a result of synergism of receptive senses which has led to bifurcation of the biological environment (environment bifurcation II in Life evolution). This evolutional metamorphosis occurred at the level of hominids, about four million years ago and resulted in the origin of *Homo perceptions*, anthropithecus. That initiated the intellectual step of Life evolution.

In terms of morpho-physiology the phenomenon of perception is nothing else but the result of the origin of qualitatively new structural elements in the brain cortex of hominids; the elements took root in its ancient layer which is responsible for instincts. Perception* allowed the anthropithecus to recreate the integral sensual images of reality, i.e. to awake qualitatively new kinds of feelings inherent in a Man only. In so doing the primary perceptive phenomena were sensual images of biological instincts, and further, in the course of intellectual evolution of anthropithecus, they were fixed genetically as relic archetypes, i.e., as structural elements of the newly formed layer of the brain cortex of *Homo perceptions*. That is why the relic archetypes may be considered as a certain psychological organ of the contemporary Man.

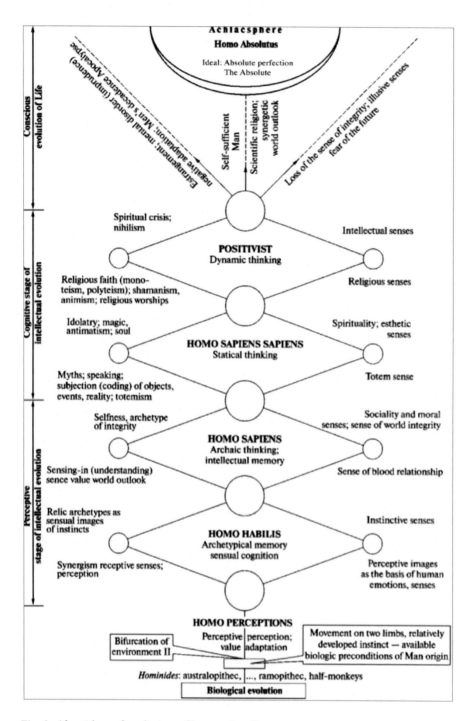

Fig. 3. Algorithm of evolution of human intellect

Thus, the relic archetypes as instinctive correlates make up the values of Man's biologic existence on the one hand, and they are the fundamental principle of senses inherent precisely to him, on the other hand. In other words, the relic archetypes serve as an intermediary between biological and intellectual steps of Life evolution.

This evolutional metamorphosis has radically changed the method of adaptation, the essence and *being* of *Homo perceptions*, since, it is owing to relic archetypes and instincts that the intentionality of his consciousness has gone up to a principally new level, and his *being* has become purposeful, active, i.e. occupational. Meanwhile a principally new method of adaptation – the value method of adaptation to the surrounding world was opened. All that became the start of formation of human psychology, that is why Man's evolution can be looked at as evolution of his psyche★ and intellect★ on the whole.

Later on, intellectual evolution of *Homo perceptions* was stimulated by the value adaptation inherent in him, which had led to the increased flexibility of relic archetypes, as well as to the enrichment and refinement of Man's senses. As a result these evolutional processes were concluded by the fact that the archetypical memory appeared under the matrix base of relic archetypes, and the origin of a new intellectual species – *Homo habilis*, paleoanthropus took place. The volume of his brain had increased essentially – it was about 1100 cm^3. He could already use primitive tools, and "sensing-in" (understanding) phenomenon became his unique quality.

Further on *Homo habilis* intellectually evolved through development and deepening of phenomenon of understanding which has led to synthesis of relic archetypes and formation of new archetypical phenomenon of Self that opened an absolutely new way of world outlook, namely, the integral understanding of objective reality by Man. It should be noted that synthesis of relic archetypes into the Self archetype took place due to the radical change of the principle of functioning of the archetypical memory. If it functioned discretely in the past in direct relation with instincts, later, as a result of evolutional development, it began functioning integrally. It served as the basis of phenomenon of self-organization of the human brain, i.e., Man's ability to think.

The occurred evolutional metamorphosis has led, in its turn, to the origin of a new intellectual species – a contemporary Man★ (neoanthropus, spiritual creature), that resulted in the adaptational bifurcation II. A unique quality of the new species, as it was noted above, was independence of his brain★, i.e., the ability to think★. Meanwhile the methods of cognition and adaptation as well as the essence of Man's *being* have radically changed. And since that time, the intellectual evolution entered its new stage, that is, the stage of cognitive development, where there occurred a leap from the sensual to rational cognition. The perception of a contemporary Man became apperceptive, and the way of adaptation and essence of *being* became spiritual instead of the value one. His psychology, which was based on the archetypical phenomenon of Self, became much more complicated.

In terms of morpho-physiology the archetypical phenomenon of Self was fixed in genetic memory of a contemporary Man in the course of evolution. This phenomenon is realized in the most external layer of the brain cortex and is of universal character common to all mankind.

Intellectual evolution of a contemporary Man lies in three hierarchic phases of development: *Homo sapiens* (totem-religious Man), Homo sapiens sapiens (religious Man), positivist Man (positivist). It is based on the archetypical phenomenon of Self, that is on spirituality, which is expressed in the sense of integrity of the world. The spiritual way of adaptation and psycho-physiological mechanism of evolution proved so efficient in reality that in terms of evolution a contemporary Man obtained by an evolutionary way such qualities as speech, art, writing, science, scientific-and-technical progress, etc., during a short period of time (forty thousand years). Owing to this the Earth's face has changed, and anthropogenic factors as to their effects and consequences reached the global scale. As a result the rate of destructiveness of Life on the Earth began to grow continuously. In this connection we should discuss the phase of positivistic evolution of a contemporary Man.

A contemporary Man always loses his spirituality because of the oppressing influence of science and scientific-technical progress; as a result his sense of the world integrity becomes dulled. He becomes a positivist with his psyche split into the consciousness and unconscious. This unordinary state of cognition radically changes the method of his adaptation, i.e., he, instead of the spiritual way of adaptation begins using the way of adaptation based on the idea of Man's supremacy over nature. And this is not a positive but negative adaptation. Since a contemporary Man losing his spirituality, also loses the spiritual way of adaptation, he leaves his evolutional way of development entering the virtual, illusory world. Precisely this adaptational metamorphosis becomes then the main cause of destructiveness of Life on the Earth.

Thus, Life which cultivated the Earth as its abode for the last several billion years obtained its master in the form of contemporary Man-positivist and has gone away from its main way of evolutional development. Positivist, as the highest phase of evolution of a contemporary Man, thinks himself to be omnipotent. In this connection he becomes a nihilist, begins to deify himself. But it is not so in reality. On the contrary, he turns into the positivist Man with one-dimensional externally controlled consciousness. His psyche is formed under the influence of positivistic view on the world, even as early as in his childhood; as a result his *being* takes an aggressive and destructive character. That is why the positivist becomes a dead-end of Man's evolution and of Life in general.

The future of mankind has faced a real dilemma: either self-insight and conscious Life evolution, or the Apocalypse. Life has passed two steps of its evolution (biological and intellectual) during 3.6 billion years, and now faces precisely this crucial dilemma. We assume that a Man, as the greatest creation of Life evolution, is able to solve this dilemma and choose his way to self-insight and conscious evolution. Also we should not forget that we are severely pressed for time to make such "realignment" happen.

The path to the conscious stage of Life evolution lies through taking principally new spirituality. In this connection let us continue the algorithm of evolution of human intellect for the future (see, Fig. 3). As was noted above, spirituality based on blind faith and religious cults has been exhausted long ago, and new spirituality is obviously expected to be oriented towards knowledge. Synergetic Theory of Life allows us to propose the scientific religion⋆ as the new spirituality common to all mankind. That is the faith in scientific knowledge that the universe is the self-organizing chaos "with no beginning or end", that is the scientific religion is the faith in that the universe is an open self-organizing chaos. And Life is a unique form of self-organization, and its evolution has led to self-insight and synergetic world outlook.

Thus, a Man, being of synergetic nature, proves to be integral not only in the world of living nature, but in the universe as a whole. That is why the scientific religion is essentially the faith in the scientific picture of the world, according to which the universe is open, integral and synergetic. In contrast to the mechanistic and physical models of the world, the synergetic picture of the universe as the nature includes a phenomenon of consciousness in a form of the phenomenon of self-organization, i.e., the consciousness as the phenomenon of self-organization becomes a fundamental property of the universe. This serves to overcome completely the incompleteness of the above mentioned models of the world.

In this connection the Big Bang theory and theory of expanding universe, currently developed in physics, is also, in our opinion, an ordinary scientific myth, since it essentially proceeds from the archaic religious and philosophic ideas of the existence of conception causes. In synergetic picture of the universe the existence of conception causes is completely denied; it is considered as the self-organizing chaos "with no beginning or end".

It should be noted that the synergetic world is not the determined, but probabilistic world. That is why the idea of transcendence of the causation principle, used in modern science, is unnecessary; it introduces respective misunderstanding in development of physical and mechanical models of the world. And what is more, the mechanistic and physical approaches to cognition of the universe do not include the phenomenon of consciousness, and as a result, they are fundamentally incomplete and non-integral. These remarks also belong completely to the Big Bang theory and the theory of expanding universe and require a principally new approach to the solution of such a significantly important question. We think that the adoption of synergetic world outlook completely solves this problem and opens new horizons of Man's evolutional development.

So, the scientific religion is the religion with no dogmas, no divine origin, no hell and heaven, no holiness, Providence and worship, i.e. it's based not on the blind faith but on knowledge of the scientific picture of the world and is both the goodwill and responsibility of a Man before Life and its continuous perfection.

The scientific religion is the new spiritual epoch with a higher level of world outlook, namely, synergetic world outlook inherent in it.

And at last, the scientific religion which synthesizes science and spirituality, i.e., knowledge and faith, opens a principally new, aimed at the archetypical ideal, absolute perfection – the Absolute of development common to all mankind, which will be the conscious stage of Life evolution.

Thus, the scientific religion is not a religion in the classical meaning of the word, not a metaphysical nor a positivistic, but a qualitatively new, so to say, synergetic view on the world. That is why Man's conscious understanding of the integrity of the world will mean the adoption by him of the scientific religion and synergetic world outlook. Owing to the scientific religion a Man will be able to understand consciously his place in the universe. Thus, a new hierarchic stage of evolution will be opened – the conscious evolution of Life.

In the future the self-insight and conscious understanding by a contemporary Man of the integrity of his existence in the universe, i.e., the adoption of the scientific religion will lead to conscious bifurcation of mankind, as a result of which a new kind of Man will appear – a self-sufficient Man*. The archetypical ideal of absolute perfection – the Absolute – will become his unique quality.

In phylogenesis a self-sufficient Man will be a superanthropus. Being a product of conscious bifurcation, the self-sufficient Man will feel from inside that he is an organic part of living nature. Besides, the phenomenon of integrity in his synergetic world outlook will include not only the world of living nature but also the universe in general.

In other words, for a self-sufficient Man the spirituality based on the blind faith as well as on different religious cults loses its power. But, at the same time he acquires spirituality and integrity at the qualitatively new level, by conscious understanding, i.e. the understanding based on the scientific knowledge of the universe, the integral world. The latter, according to STLifeife, is synergetic, open and integral, while a Man is only its indivisible part.

A self-sufficient Man as a new kind of Man will be the master of his own destiny. First of all, a Man will repudiate the priority of the value world created by him in favor of the really existing world. In so doing the Life evolution acquires its purpose and meaning in his person, while a Man finds his present; he will get the purpose and the meaning of his existence not from the past, but from the future. The Man's *being* will be oriented towards the archetypical ideal, to absolute perfection – the Absolute.

It is natural that the conscious bifurcation will also be embodied in other specific peculiarities of a self-sufficient Man. In particular, the world will be not only open, synergetic, but also integral in his world outlook. Moreover, the subject-object division of the world currently accepted in modern science and philosophy will lose its meaning and become fiction for him, since in a synergetic world outlook the subject merges with the object. In this connection the intellectual world of the self-sufficient Man will radically change. A new syncretic (integral) science★ will substitute the present-day objectivistic science. It will be based on the synergetic view on the world. Cognitive processes in it will be based on scientific intuition (ingenuous striving), scientific-technical and technological creativity. The pragmatic aspects of scientific and scientific-technical problems, i.e., development of algorithms of human activity on the improvement of Life quality will become the priority aim of the syncretic science, rather than fundamental-theoretical investigations directed to world outlook.

The present education system will be radically changed. The humanized education★, i.e., the education system based on synergetic view on the world and on scientific religion will take its place. The study of human nature will occupy a central place in this system. The absolute perfection, the Absolute, will become an ideal of the humanized education; it opens the way to self-insight, i.e., helps a Man to become a self-sufficient one. In other words, students in the system of humanized education will learn to create their future. That is why the preparation and education of a self-sufficient Man will be a final objective of this education, in contrast to the present education systems.

As a result of self-insight the self-sufficient Man will acquire a principally new attitude towards himself. First of all, he will change the conception of health. The present directive therapy will be changed by individual (or psychosomatically integral) therapy★. It will consider any pathology as a psychosomatic manifestation, that is, as the integrally organismic phenomenon.

Since, according to STLifeife, the consciousness and body are always syncretic (merged, integrated), the division of diseases into somatic and psychiatric is conditional and has no scientific basis. The division of a Man into a body and consciousness is a myth which has nothing to do with reality. Disorders in psyche and nervous system under corporal sufferings, and vice versa, appearance of different somatic deviations under the effect of psyche and related nervous factors are rather natural. There even exists a certain notion, that the Man's physical appearance symbolically reflects the state of his mind. The nervous system serves as a connecting element (intermediary) between the somatic and psychic diseases. Therefore, its diagnosis and evaluation of its state will be of the key value in individual therapy. More simply, health is the integrity of Man's organism, i.e., the integrity of Man's soul and body. Thus, the psychosomatic integrity as a basic principle of the health conception will make up the basis of individual therapy, which will become the beginning of a new stage of development of medicine.

The other fundamental property of a self-sufficient Man will be his archetypical ideal, absolute perfection – the Absolute as evolutional sublimation of the archetypical phenomenon of Self. The ideal is not simply a product of rational comprehension, but the archetypical ideal which determines the *being* of a self-sufficient Man. It will appear as a result of conscious bifurcation and will be fixed in mentality common to all mankind. Thus a self-sufficient Man's *being* in contrast to that of a contemporary Man will be directed to the absolute perfection – the Absolute. Of course, this evolutional metamorphosis will radically change the psychology of a self-sufficient Man, since it is the ideal that will become its centre, nucleus rather than the previous archetypical phenomenon of Self, which is a symbol of Man's understanding of the integrity of the world. And this psychology of the absolutely new kind can be called meta- or enlightened psychology. It will be based on the knowledge by self-sufficient Man of the scientific picture of the universe and on his understanding of his own place in it, i.e., his adoption of the scientific religion. And this means, that the enlightened psychology will not be hidden and unyielding to rational cognition anymore, but it will be clear and indivisible. On the whole the self-sufficient Man will strive in his evolution to *Homo absolutus*★ – a perfect Man, as a result of which the biosphere will be sublimated into the Akhlaqsphere★.

Therefore, mankind will not necessarily find its future in the Apocalypse, for it can also be found in the Akhlaqsphere, that is, in the sphere of dominance of scientific religion, synergetic world outlook and moral-ethic values common to all mankind.

The author wishes whole-heartedly that a Man would realize all that as soon as possible and would take greater responsibility for his future. A Man has to realize that the present forms of social order are extremely primitive and catastrophic. Thus the future will belong to a qualitatively new social order, a mature society★ that is the society of self-sufficient people (individuals), with an inherent synergetic world outlook and scientific religion.

Conclusion

Let us make a scientific forecast for the nearest future of mankind. With this in mind, we'll first note some fundamental aspects of the Synergetic Theory of Life. According to it Life on Earth was not created, it was not brought from the outside and it was not a result of discovering the secret essences, but is a phenomenon of self-organization based on synergism of high-molecular compounds (polymers) – proteins and nucleic acids. Its appearance led to bifurcation of physicochemical environment which existed on Earth about 3.5 billion years ago and it has synergetic nature. Self-reproduction and adaptability are unique qualities of Life.

The Life evolution is of adaptation-bifurcation nature and consists of the following levels:

Biological level (cellular or metabolic evolution, genetic or morpho-physiological evolution) **intellectual level** (archetypical evolution or psycho-physiological evolution) – – **conscious level of evolution** (self-insight, adoption of scientific religion and formation of synergetic world outlook).

An interrupted arrow in this diagram indicates the future. This canon, or algorithm of Life evolution, includes not only the past and the present of the phenomenon of Life but also its future. The highest hierarchic level of Life evolution is a contemporary Man, a spiritual creature whose behaviour is determined by his individual system of values and relic archetypes, representing the values of his biological existence.

From the view point of evolution Man, as a unique biological species had existed for about 40,000 years. And the Life history of each individual, including his embryonic development, contains a full algorithm of Life evolution. According to STLife, a Man is in evolutional crisis now, which is why the Apocalypse presentiment is wandering around the globe. Since a contemporary Man in the phase of positivist, losing his spirituality, is in evolution crisis, i.e., in search for new spirituality, he absolutizes, and even deifies himself. However he does not have a corresponding way to appease his nihilism, narcissism and unrestrained money-grabbing which are most brightly displayed in his estrangement and thirst for consumption and pleasure. And this takes baneful influence on his *being*, i.e., as a result of the evolutional crisis of a contemporary Man his *being* becomes more and more disastrous and his social life is subjected to chaos. A general tendency is such that both technics and technology make progress while Man is regressing. As a result, firstly, the Earth's resources become even more depleted and Man's domination over nature is stimulated which is a direct threat to the existence of Life as a phenomenon of unity and integrity. Secondly, social contradictions are intensified at all levels: from the family life to mankind's existence. Today it is most clearly manifested in Man's moral degradation, in separatism, terrorism (from individual and group to government one), vandalism and other forms of barbarism. Thirdly, a probability of technical and technological catastrophes, including possibilities of using the achievements of scientific-technological progress against Life, increases. All this sharpens a feeling of inevitability and gives rise to a principally new kind of fear, the fear of the future.

Thus a contemporary Man, using a negative way of adaptation – the adaptation by domination over nature, has completely lost his primordial evolutional way of development and is in the dead-end of Life.

General probability of destructiveness of Life is a sum of probabilities of each factor noted in *Introduction* and maybe other, including random, factors of manifestation of contemporary Man's evolutional crisis. Today it is this probability that determines the psychological "landscape" of Life on Earth. And this "landscape" is very dynamic and unstable. However, it is evident that quantity, rate, and diversity of destructiveness of Life are persistently growing, and it is thought this growth will be nonlinear, that is, the destructiveness after certain time will be developing by the pattern of chain reaction and Life self-destruction will be inevitable. Figuratively speaking, a certain destructive "explosion" of the living nature itself is expected in a due course of time. Scientific intuition of the author allows him to say that, with the maintenance of Man's present credo and way of existence deprived of evolutionally necessary conscious bifurcation, the time limit of the approach of this phenomenon, the phenomenon of inevitability of Life self-destruction, is no more than one hundred and fifty years. This time may be much less in reality. I think that in this historically short period Man will be successful in comprehending the nature of Life, nature of himself and in finding the way to synergetic world outlook and principally new type of evolutional development that is the conscious evolution of Life. The latter, as the result of conscious bifurcation of humanity, will be opening a qualitatively new way to evolutional development of Life, and its unique quality will be the appearance of a new Man – a self-sufficient Man, superanthropus. He will have a synergetic view on the world, and his *being* and behaviour will be determined by the archetypical ideal – the striving to the absolute perfection, the Absolute. And only under such conditions Man's intellect and his *being* will be not destructive but creative and constructive. Such a sublimation of intellectual evolution into conscious evolution is the only way out of the evolutional crisis of a contemporary

A Man as the crown of Life's evolution should not lose this salvational chance. The author wants to believe in that and emphasizes that the future of mankind is in its own hands: in its self-insight and in its conscious evolution. There is just no other way!

The author wants to express his sincere gratitude to his wife Ogulsapar, daughters Ogultach and Gozel, sons Ata and Batyr for their constant support without which this book could have stayed unfinished.

Explanatory Dictionary of Terms
of Synergetic Theory Life

Relic archetypes (instinctive correlates) – emotionally-perceptive images of biological instincts which resulted in the origin of qualitatively new species of hominids, *Homo perceptions*, anthropithecus and bifurcation of biological environment, that is the environment bifurcation II. In due time it was this that marked the beginning of intellectual stage of evolution of Life. Thus, instincts are the biological roots of archetypes.

Archetypes, including relic archetypes, which are emotionally charged, i.e. "alive", reproduce non-discursive symbols that belong to human values. By contrast, the discursive symbols belong to the sphere of facts and rationality. In the course of Life's evolution the *relic archetypes* were being fixed in the genetic memory of *Homo perceptions*, anthropithecus, and proved to be a connecting link between the biologic and intellectual stages of Life evolution. In other words, they are unified episodes of former lives and are of universal character, common to all mankind (collective unconscious, according to C. Young).

The *relic archetypes* are expressed in neoanthropus – a contemporary Man – in his dreams as archetypical symbols and express the world of his values of biologic existence. It is of interest to note that the intellectual level of a contemporary Man in early childhood seems to resemble that of archeoanthropus, a relic Man, in his final stage of evolution, i.e. a transitional Man in the person of the Neanderthal man. A comparative analysis of the psyche of a contemporary Man in his early childhood with the psyche of people who become wild in their childhood can be invaluable in studying this phenomenon. They do not speak and are still syncretic with the environment, and their behavior is mainly determined by *relic archetypes*, representing the instinctive correlates.

It should be especially noted that the relic archetypes are an initial element in a complex chain of archetypical phenomena, which serves as a bioenergetic channel of psyche of a contemporary Man. Since archetypes are both the images and emotion, and an emotionally charged image becomes sacral (numinous), i.e. acquires psychological energy, one can suppose that emotions are carriers of Life energy. Besides, the *relic archetypes* as instinctive correlates give the human consciousness an attribute of intentionality (orientation of consciousness towards the object) and form his value attitude towards the surrounding reality, as well as determine his will. In other words, it is the archetypical phenomena that determine the conscious determination of a Man and appearance of new senses peculiar to him only, i.e. human senses. As a result of these evolutionary changes, in contrast to the actions of other living creatures, human actions acquire the active, occupational character.

In terms of morpho-physiology (somatic) the *relic archetypes* are sensory images of biological instincts, they are fixed in the genetic memory of a contemporary Man.

Akhlaqsphere – a sphere of domination of scientific religion, synergetic world outlook and moral–ethical values common to all mankind. Social life in the *akhlaqsphere* goes on in a single spiritual space and directed towards archetypical ideal – the absolute perfection, the Absolute.

Homo absolutus – a perfect individual, conscious of the integrity of his existence in the world and striving for archetypical ideal – absolute perfection, the Absolute – a self-sufficient Man. As to religious canons he will be the enlightened Man, who has overcome himself and has become a microcosmos as an image of the universe. The world of Man's *being* will no longer be a world of illusory values, i.e. not a world of religious symbols and archetypical imperatives of *being* created by him, but an actually existing world.

Spirituality and spiritual way of adaptation. *Spirituality* in a broad sense is a sublimation of archetypical phenomenon of Self in the consciousness of a contemporary The latter expresses the understanding by a Man of the integrity of his existence in the world. That is practically realized both through the ethical and behavioural imperatives of *being* (Zoroastrism, Daocism, Buddhism, Sintoism, Jainism, Confucianism) as well as through the blind faith (Judaism, Christianity and Islam). That is why *spirituality in general understanding* – is the way of life of a contemporary Man, based on his sense of integrity of existence in the world. *Spirituality* includes not only the image of the present, but also the image of the future, that is the image of eternal life, immortality. By losing spirituality, a Man loses not only a sense of the integrity of the world but also the image of eternal life. In this connection, by being subjected to the internal crisis, that is fear of the future, he degrades.

Therefore, the loss of *religious spirituality* is the degradation of Man's ability to understand the universe.

Awaking different images and myths concerning the integrity of Man's *being* in the human consciousness, spirituality, as an archetypical phenomenon seizes him, carries him away and creates necessary conditions or the basis of his adaptation to the surrounding world. This principally new kind of adaptation peculiar only to a contemporary Man may be called a spiritual way of adaptation to the surrounding world. An intelligent world outlook by means of deduction by analogy serves as its basis. Hence, it follows that under the lack of spirituality neoanthropus – the contemporary Man, losing his adaptability, will naturally suffer from the evolutional crisis. This circumstance radically changes the existing notions regarding the genesis of religions and their significance.

Spirituality as a religious world outlook is the greatest discovery of the evolution of Life, it cannot be considered as the remnant of history or as a certain component of public consciousness. One should never forget that religion is the basis of culture (civilization), and evolution of a contemporary

Man is determined by *religious spirituality*, which is a necessary condition of his adaptation to the surrounding world. In a word, a contemporary Man cannot do without religion, without spirituality at all. And he, as a spiritual creature, takes any action made against his spirituality as blasphemy and an attempt on his own life and the lives of his nearest and dearest.

Such phenomena as mysticism, myths, magic and idolatry (paganism), shamanism are directly connected with the origin, existence and loss of spirituality. They are expressed in the modern conditions as unordinary states of consciousness and as spiritual crisis. That is a characteristic feature of a contemporary Man who feels the awakening and renewal of *spirituality*. But if mysticism and myths are characteristic of the stage of origin and functioning of *spirituality*, magic is characteristic of the stage of its loss. Hence, it should be emphasized, that neoidolatry as the worship of the wealth of scientific-technical progress is caused by degradation of the existing *spirituality*. But in reality this phenomenon is often taken by mistake as remythologization. Indeed, because the religious myths are being dethroned, a lot of new ersatz spiritualities are generated (religious sects, associations). An analogous phenomenon is observed in the contemporary objectivistic science, since being more and more differentiated it is more and more mystified. Thus, I think it is very important to reveal the essence of the phenomena of magic and shamanism.

Magic is the sorcery, wizardry, created by Man by synthesizing knowledge and art. It, "turbulizing" the integral mythological world, exerts a negative, corrupting influence on his *religious spirituality*. Thus, magic, essentially, is the belief in Man's ability to exert supernatural influence on people and natural occurrences.

Shamanism is a relic religion that creates the integrity of Man's existence in the world. Such integrity (or ecstatic state of the shaman) is created through associating the shaman's soul with spirits (ghosts), demons and other illusory anthropomorphic creatures, which are the products of mystic searches; i.e. the shaman, inhabiting the real world with mystic beings, creates the integrity of Man's existence in the world. In so doing he makes the best use of magic "technique" and serves as an intermediary between this mystified world and the real world. Then a new quality appears and develops in a Man through such an illusory construction; that is the sense of space, and he rises from the level of totem thinking to the level of static (religious) thinking. In the end all this increases the adaptability of a group of people, taking part in shaman's rite, to various extreme life conditions (disease, drought, famine and other phenomena which threaten Man's existence).

Spirituality, being a product of archetypical phenomenon of Self, puts in order the world chaos using myths, and thus, creates a necessary condition for Man's adaptation in real world. This is exactly why the *religious spirituality* is of absolute importance in the individual system of values of a contemporary Man, and *intellectual spirituality* (philosophy, literature, science, etc.) – is a derivative, it is secondary and is not connected directly with archetypical phenomenon of Self. They proceed from Man's *being* which is deprived of primary spirituality. Thus, the historical rather than evolutional act of religion's secularization from state and school, implemented in Europe, has led to scientific-technical, technological progress on the one hand, and has accelerated the loss of spirituality by Man on the other hand. As a result, currently, when technics and technology are progressing, a Man, on the contrary, is continuously regressing, losing his spiritual principle, that is, he is degenerating into nihilist and positivist, and his decadence takes place. In other words the human society is transformed into "technological society".

Europeans with their nihilistic act have, apparently, thrown away spirituality. The echo of this historical fact based on the erroneous ancient Greek study about *Homo sapiens* – an intelligent Man, is displayed today, with all clearness and obviousness, in life of western civilization and in modern history in general. For it is such an erroneous conception of a Man that has become a cause of European nihilism – attitudes which negate traditions, spiritual principles of *being* and other principles of social life. The consequences of this historical act will be considered in more detail further on.

Thus, the synergetic theory of Life does not only not negate the idea of fideism and neo-Thomism about primacy of religious faith over rational knowledge but also reveals their scientific meaning. In contrast to the existing notion it permits making a conclusion that only *religious spirituality* and spiritual way of adaptation stimulate a Man to become even more perfect and civilized. The formation, development and degradation of *religious spirituality* happens, following the law of "three M", i.e. by a sequence

$$\text{mystics} \longrightarrow \text{myth} \longrightarrow \text{magic.}$$

If ordinary metabolic and morpho-physiological (genetic) mechanisms of evolution and adaptation of living creatures are slow evolutional processes and have practically exhausted themselves at the human level of Life, then owing to spiritual way of adaptation neoanthropus – a contemporary Man, during a relatively short period of time (30,000 – 40,000 years) has discovered not only his intellect through the way of evolution, but also such fundamental phenomena, as language, art, philosophy and rational cognition. Owing to them, he initiated an active and broad-scale adaptation of nature to his needs. Figuratively speaking, the phenomenon of adaptation has changed its "sign" that is a Man began adapting to the surrounding conditions by fitting the latter to his own needs. In other words, a contemporary Man, by swapping his own natural spiritual way of adaptation with that based on his predominance over nature, deviated from his way of evolutional development. And this, in its turn, awakened and fastened his positivistic attitude to the world around. Meanwhile science by moving further away from spirituality becomes a potentially dangerous instrument of the adaptation of the environment to the illusory needs of a Man.

Science, by alienating from nature i.e. detaching from reality, becomes its destructor. This is exactly why ecological, technological and many other kinds of crises are becoming more and more intensified. One should think that this process has achieved a critical scale, and thus it is of a catastrophic character. But the main thing is that a contemporary Man himself degrades because of the loss of spirituality and, respectively, collapse of adaptability. In search of new spirituality he is strongly subjected to the influence of different socio-political trends and illusory intellectual, scientific-technical, ideological, political ideas and imperatives.

Moreover, it is because of the spiritual crisis that a contemporary Man suffers from excessively stupefying influence of material values and riches in general. Thus, his attitude towards the world around becomes more and more mercantile. It should be noted that the current tendency to integrate, that is globalization, is unfortunately based not on spirituality which is adequate to the nature of a contemporary Man, but on the technical and technological intellect which opens the way to technosphere. And that is a road to the dead-end of mankind's development, since globalization and other forms of economic integration – such as economic coalitions (EEC, OPEC, ACEAN, etc), transnational companies (BP), when developing economy and technology, generate elitism, which in its turn stimulates not the social integration, but vice versa, intensifies social disintegration. They destroy *religious spirituality*, fundamentals of moral-ethical norms, traditions and, by developing separatism, "turbulize" the society.

Jesus Christ said before the crucifixion: "Father, forgive them, for they don't know what they do". Indeed, a contemporary Man does not know the nature of Life and his own purpose, and he does not understand that without changing his system of values, without changing his behaviour, his treatment of people and environment, he dooms himself to decadence and dooms Life itself to death. That is because the lack of spirituality, while degrading Man's understanding of integrity of the world, deprives him of the main attribute given by nature – the spiritual way of adaptation.

The spiritual crisis and the lack of spirituality which are inevitable consequences of intellectual evolution have completely revealed the problem of a soul. In ancient time, Man, who had found some nonmaterial essence besides the material basis in himself, made the greatest discovery: he opened the existence of his own soul, which is a mythologized metaphor and has no material basis. Later in process of intellectual evolution he began ascribing this quality of his to the world around and thus animated it (animatism). Such mythological notion created the basis for the formation of *spirituality*, the integral comprehension of the world by Man. Animatism had soon developed into animism which assumed the existence of spiritual (nonmaterial) creatures that are presumably able to communicate with a Man. Today, animism is an independent structural element of all religions.

Thus, a soul and *religious spirituality* normally are always syncretic and directly connected with Man's intellectual evolution. That is why in the contemporary life, when a Man is in spiritual crisis his soul also becomes demonic, destructive. To avoid this unordinary state of consciousness, the notion of "soul" is always depreciated and its place is taken by the new notion of "psyche". But such change of notions has not solved the appeared problem, since it soon became clear that the contemporary Man's psyche, in the general case, was not integral and unambiguous, that is, it has in reality two aspects – conscious and unconscious. The fact is that Self (archetype of integrity), being a nucleus of psyche in the consciousness of contemporary Man, manifests itself as *religious spirituality*. So, the phenomenon of the loss of spirituality for Man means both a spiritual crisis and a loss of the sense of integrity of the world. From this point of view, the soul is the integral psyche peculiar to a spiritual Man, while split psyche is inherent in a contemporary Man – a positivist. In other words the soul is archaism for the positivist. Thus, from the viewpoint of evolution a genuine contemporary Man is a positivistic Man, and it is he who being at the top of intellectual evolution becomes the deadlock of Life.

The way out of the crisis situation by the mechanical change of the "metaphor" of soul by a more extended notion of "psyche" has now become inefficient and incorrect. This problem proved to be much more complicated than expected. A qualitatively new kind of *spirituality* is required, which would be able to awaken a sense of the integrity of the world in a Man, and provide his further spiritual growth and evolutional development.

Today the renewal of *spirituality* is a fortune-bearing problem of all mankind, since it is impossible without it to stop the process of further sharpening of such catastrophic deviations as spiritual crisis, psyche splitting, illusory perception, mental disorder, deadaptation and at last to avoid decadence of a contemporary Man. It may be noted as an example, that there appears quite clearly the image of technocratic development and construction of the society of one-fifth, that is the elite society where eighty percent of the global population is under the threat of unemployment. They are considered unnecessary people. And that is a bias towards eugenics, Malthusianism and social-Darwinism, that is orientation towards the artificial reduction of population and selection of a new breed of people. Such technocratic way towards globalization is directed towards the scientific-technical progress rather than to spiritual growth of a Man, and is dangerous on its own. This concept of globalization bears a remote resemblance to the communist slogan "Workers (Proletarians) of all countries unite!". But proletarians are supposed to be substituted by intellectuals. All this is the evidence of the fact that the arena of Life's evolution has now transferred to the sphere of global problems that is those common to all mankind.

An abyss has appeared between financial and commodity markets, and the world economic system is being destroyed. The world is ruled more and more by multinational corporations and industrial-financial elite rather than by the sovereign states, or their union. Under the lack of spirituality, side by side with the destruction of natural environment and aggravation of social and political contradictions the contemporary Man himself, being subject of mental disorders, is in the deadlock of Life evolution. And that is the eschatological meaning of our epoch.

Life – a phenomenon of self-organization based on synergism of high-molecular compounds: proteins and nucleic acids. The origin of Life has led to bifurcation of physico-chemical environment which existed on Earth about 3.5 billion years ago. Self-reproduction and adaptability of Life being the basis of its evolution are its unique qualities. That is why Life, in its essence, is evolution, i.e. the living nature can exist only under conditions of evolutional development.

The initial form of Life was, apparently, much more primitive, than the simplest one-cell organisms of today. In terms of structure they may be considered as protocells. And in nomothetic context the synergism of nucleic acids and proteins may be considered as a principle of uniqueness and unity of *Life*. Indeed, it is because of synergetic nature of *Life* that all living organisms on the one hand have a single genetic code, and, on the other hand – each of them is individual, that is, there are no two absolutely identical cells in nature, to say nothing of the identical living organisms. At the human level of Life existence this principle is most clearly displayed as the phenomenon of Man's duality, i.e., he is both individual and social, which is why all existing social, philosophic, economic conceptions and theories based only on the unilateral approach to Man and his *being* are either primitive or absolutely unacceptable. It is often assumed that the difference between social and gregarious forms of existence is of quantitative rather than qualitative character, i.e. a Man is more intellectual than other living creatures. But their difference does not lie only in the degree of intellectual development, it is of principle, since, if the herd is based on instinct, sociality is based on spirituality. That is why the loss of spirituality by Man leads to "turbulization" of society, i.e. to social chaos and global catastrophe.

Life evolves following the adaptation-bifurcational principle, i.e., if evolution on biological level happens genetically through the appearance of a new biologic species, then, at Man's level it happens through his psyche. *Life* evolution itself is of dual character: it is both gradual, and leap-like. As to the current evolutional crisis of *Life*, it is a result of decadence of a contemporary Man who has lost his spiritual basis. The future of *Life* will be determined by conscious bifurcation and appearance of a new species – self-sufficient, and in religious terms enlightened Man (individual).

In connection with the above mentioned it should be noted, that the existing theory of evolution by C. Darwin relates to only one stage of *Life* evolution. It is not based on the nature of *Life* in general and on Man's nature in particular. It needs full reevaluation, since the biologic evolution in its essence, is only a fragment of Life evolution. And its mechanism is not a biological selection, but somatic development and perfection of living organisms.

One should think that Darwinism and its central notion – biological selection, are long out-of-date and today are of interest only from the viewpoint of the history of science. As to the attempts to adapt scientific achievements of the 20th century, in particular the achievements of genetic science, inheritance theory, molecular biology to already obsolete evolutional theory and creation of the synthetic theory of evolution – they are waste of time and resources. We need a qualitatively new level of world outlook and more developed and perfect theory of Life that has been done, in our opinion, in this given work.

Individual (psychosomatically integral) therapy is the therapy based on psychosomatic integrity of organism functioning, being created by an individual system of values which takes roots in relic archetypes responsible for senses peculiar to Man only. The individual system of values has three levels: the highest level in the norm – religious-spiritual one (including religious worships and everyday life imperatives), proceeding from the nature of a contemporary Man as a spiritual creature. Then follows the level of totem and moral-ethical values, and finally the last one – the level of cultural values. They are not interchangeable, but closely interacting with each other, create a dynamic world of Man's values which determines Man's consciousness, psyche and *being*. These values, when interacting and opposing, create a complex mosaic of life images. That is why the functioning of the Man's brain, his consciousness and psyche are subject to various forms of disorders, including a spiritual crisis, psychosis, paranormal psychological phenomena, etc. But these, essentially, reversible abnormal changes should not be mixed with pathologies connected, in their nature, with genetic, biochemical, in a word not functional, but irreversible organic deviations.

The individual system of values, when modeling the value world of Man or creating his world of symbols, is expressed first of all, in his temperament, in mentality and determines in many aspects his will, psyche and self-realization i.e. his establishment and destiny. And the same system in normal state, as was mentioned above, creates the organismic (psychosomatic) integrity of a contemporary Man, since there is every reason to think that numerous somatic diseases are based on the nervous-psychic problems, which arise as a result of the stress in the integral world of Man. Only genetic and organic cases of diseases can be an exception here. That is why the diagnostics and evaluation of the state of the individual system of values become the necessary conditions of efficiency of the somatic treatment. In psychoanalysis it is thought that neurosis is caused by sexual drive forced out of consciousness into unconscious. One should think that it is a long out-of-date and essentially erroneous notion of neurosis, since there is no doubt that neurosis is also directly connected with stresses arising in the individual system of values of Man.

Thus, the somatic treatment is most efficient only when it is based on evaluation of the state of Man as an organismic integrity and on analysis of his individual system of values. Besides, a correct structuring, in particular the removal of ersatz values which have occupied the place of spiritual, moral-ethical values, that determine the Man's quality of Life, is necessary not only for a patient, but for somatically healthy Man as well in his establishment and self-perfection. And it should be emphasized, on the whole, that the process of mutual devaluation and overvaluation of values is a fundamental peculiarity of psyche of a contemporary Man and it looks impossible to unambiguously evaluate and diagnose his pathology without bringing it to account.

The process of substitution of religious spirituality with surrogates is the most tragic and dangerous for the health of a contemporary Man, since this process destroying the structure of the individual system of values and, respectively, the psychosomatic (organismic) integrity, generates neurosis, schizophrenia and other nervous–psychic disorders. By the way, when we say schizophrenia we mean a phenomenon of devaluation of Man's religious-spiritual values and their substitution with illusory values. A phenomenon of psyche split into the consciousness and unconscious is considered a result of constant inhibition of religious spirituality by rational cognition, technical intellect and negative life experience.

These and other analogous anomalies in the functioning of the contemporary Man's psyche give birth to illusions and erroneous perception of reality by him; as a result his behavior becomes de-adaptive, pathological and even fatal. Thus, the *psychosomatic integral therapy* which considers a Man as a spiritual creature will become not only more efficient but the principal instrument of the upcoming conscious evolution, as well as it will become a basis of education system and self-perfection of a Man, that is the ABC of humanized education, study of human nature in general. And the reference point will be the absolute perfection – the Absolute, as a symbol of Man's perfection. All that means that in the future, owing to *individual therapy,* a contemporary Man, understanding his responsibility for preservation of his own health and a necessity of continuous self-perfection, will be able to create himself and to become self-sufficient (an individual).

Thus, in contrast to the current directive medicine, that accentuates the pathology of organic (somatic) origin, the *individual therapy* will consider any pathology as psychosomatic that is organismically integral phenomenon. Since, according to the synergetic theory of *Life,* the consciousness and the body are always syncretic (integrated), and division of human diseases into somatic and psychic has no scientific basis.

The division of a Man into body and consciousness is an erroneous religious and philosophic notion, having nothing in common with reality. The disorders in psyche and nervous system under physical sufferings and vice versa, the appearance of various somatic deviations under the effect of psychic and related nervous factors are also rather natural.

There even exists a certain notion that the physical image of a Man reflects his psyche. The central nervous system (head and spinal brain) is a connecting element, a "bridge", between somatic and psychic diseases. One should not forget that nerve disease in a contemporary Man mainly develops against a background of neurosis. That is why the diagnostics and evaluation of its state in *individual therapy* will be of key importance. Simplistically, health is psychosomatic balance and organismic integrity of a Man, i.e. the integrity of his body and soul.

Therefore, psychosomatic (psycho-physiologic) balance and organismic integrity as a basic principle of health conception will be the basis of *individual therapy* and become the beginning of a new stage of development of medicine. In particular, all problems of contemporary medicine, connected with the erroneous Old Greek conception that a Man is an intellectual creature, will be automatically completely taken off. As is known, there is no place for religious spirituality and spiritual crisis in the mentioned conception. In contrast, the *individual therapy* proceeds from the fact that a contemporary Man is a spiritual creature, and that is why the psychosomatic integrity is a necessary condition of his existence. In other words, if the contemporary medicine is based on the conception that Man is an intellectual creature, then the *individual therapy* will be oriented to the conception that he is a spiritual creature. And this will help to completely overcome the one-sidedness of modern medicine, and new opportunities in Man's self-treatment and self-perfection will be opened.

Intellect, in a broad sense, is the cognitive ability of living creatures, being realized through sensations, instincts, sensual cognition and thinking; it is the method of their *being*. In a narrow sense it is "sensing-in" (understanding) and cognitive ability, that is the ability of sensory and intelligent comprehension of the world inherent only in neoanthropus, a contemporary Man.

Mature society (civil society). According to synergetic theory of Life, as a result of self-insight and adoption of scientific religion, the appearance of a new species – a self-sufficient Man is expected in the nearest future. The above said in its turn will lead to mankind's bifurcation (split into two parts), one of which is self-realized and another one is not self-realized. We understand the *mature society* as the community of those, self-realized, self-sufficient people (individuals). The synergetic world outlook and a single spiritual space that is the scientific religion and common archetypical ideal, absolute perfection – the Absolute, will be peculiar to them.

It is after self-insight and comprehension of the world integrity by a Man, that his archetypical phenomenon of Self is sublimated into archetypical ideal, the absolute perfection, the Absolute.

The mature society will substitute all the already existing non-mature ones that are based on the primitive philosophical, ideological, economical and political myths of the society that absolutize individualism or collectivism, democratism or totalitarism, legal or patriarchal, and other kinds of social order analogous to them, which do not account for the primacy of a Man and his evolutional canons over so-called objective laws of social development which do not exist in reality. The projects widely disseminated in the western world and catastrophic in essence, such as the theory of convergence, the idea of globalization, the society of one-fifth and other analogous approaches to social development will be also removed.

Opportunities of empirical sociology, the significance of technical intellect and scientific-technical progress are overestimated in all the above innovations to the detriment of spirituality and humanization of socio-economic life of the society. That is, in essence, the way to the technosphere, the social anarchy i.e. "turbulization" of society, being characteristic of it. This absolutization of technics and technical progress for a Man is a symptom of neobarbarism, i.e. technological barbarism and the approaching Life dead-end in general.

One should remember that technical progress in the hands of an immature Man can prove to be a real catastrophe, threatening the Life's existence on the Earth. To put it differently, the way to technosphere under lack of spirituality is the way to self-annihilation. Thus the Man should not bow down before technical progress and its wealth to preserve Life on the Earth. He must "override" it and turn it into a conscious constructive force. And the road lies through the scientific religion as a qualitatively new level of spirituality.

According to synergetic theory of Life a Man is both individual and social, and his evolution is determined not by the rules of linear and progressive development but by canons of Life evolution, i.e. according to the adaptation-bifurcational pattern. That is why the widely used substitution of these canons by some social myths and ideas, based on various ersatz imperatives of *being*, is a dangerous delusion of a contemporary These myths are based either on primitive and erroneous Man's conceptions and ideas, or on a simple mechanistic extrapolation of some particular positive qualities of the existing social orders into the future.

It is natural that social, scientific-technical, cultural, economical, political factors in a similar way to the great natural phenomena, including planetary ones, such as the origin of the Earth's secondary atmosphere, change of climate, continental drift, solar-Earth and space anomalies, etc., can both accelerate and decelerate, and maybe even halt the evolutional process, in dependence to their phase relations. But the understanding of primacy of Man's evolution and Life evolution in general over such natural and humanitarian factors is a necessary condition of maturity of the society. In particular, the historical act of religion's secularization from the state and school, implemented in Europe, considerably accelerated the approach of today's evolutional crisis of a contemporary Man.

Thus, in the future in *mature society* a new civilization common to all mankind, freeing itself from social remnants, myths and estranged from reality catastrophic innovations, will live by the canons of conscious evolution, i.e., evolution in which a complete syncreticity of a Man and nature, Man and universe in general will be achieved. That will be really a daoistic way of Life. Under these conditions the biosphere will be sublimated neither in the noosphere nor technosphere, as is often noticed in scientific and popular scientific literature, but in Akhlaqsphere*, that is the sphere of domination of archetypical ideal (absolute perfection, the Absolute), scientific religion and moral-ethical values common to all mankind. This will also completely and finally overcome such social diseases as spiritual crisis, lack of spirituality, loss or collapse of adaptability of a contemporary Man and his decadence. But the main thing is that a Man in mature society will not be a blind puppet of illusory laws of social development anymore, but he will live following natural canons of evolution of Life itself. Only under these circumstances a Man will stand over the politics, over the state, while a society will be a civil one in the full meaning of the word, that is the community who became self-sufficient (individuals). Thus the power of law itself without corresponding spiritual and ethical basis is not enough for a Man to become an individual, and society – a civil one.

I think that the road to civil society lies only through a new conscious bifurcation and origin of a new species, a self-sufficient Man (true individual), rather than through simple, so-called democratic and legal transformations of the society as is customary in the existing socio-political notions. The matter is that the contemporary Man's *being* is not subject to logic, and that is why it is always outside the limits of legal space. In this connection the vitally important work lies ahead – that is the work on rethinking the existing socio-political notions, notions such as "individualism", "egoism", "alienation", "people mass", "party", "politics", "state", "democracy", "social system", "philosophic system", "cosmopolitan", "proletariat", "marginals", etc.

It should be noticed the democratization from top to bottom is the evolutional way of development of the society, while democratization from bottom to top is a revolutionary way of development. As to their nature they are synergetically interrelated and thus should balance each other, that is especially necessary for democratic reforms in economy and politics. An indisputable condition of success of any democratic reform is availability of the idea of national culture, its source being in the mentality, history of each people and socio-political values common to all mankind. However today most people understand democratization as the one from bottom to top. More simply, democratization is often understood as Americanization, that is individualization and taking the American democratic values as a standard for all, and American way to democracy is considered universal. I think this is not the case though, since the American history is completely different to the history of other peoples. It virtually has no historicity, traditionalistic load, etc., that is why the mentality of Americans mostly does not proceed from traditions and religious morals, but is based on law.

In *mature society* freedom and democracy will be conscious and based on scientific religion and moral and ethical norms common to all mankind. It should be noted that the ideas of "freedom" and "democracy" in essence are ambiguous, abstract socio-political notions. In particular, the democratic field is so wide that its one side borders with anarchy and the other with authoritarism. One can say that everybody understands democracy in their own way. But in practice democracy is realized in specific historical and socio-political conditions in specific mental, moral–ethical and legal space. Beyond them it becomes ambiguous and loses its constructive force.

Export of democracy without accounting for the existing natural–historical values, level of development of consciousness and mentality is a questionable venture which can be even pernicious. Thus the constructive freedom and democracy should emanate first of all from mentality and be oriented to the values common to all mankind (right for life, freedom of conscience, freedom of speech, equal rights, justice, etc.). There arises a new important question, how the problem of variety of cultures will be solved in *mature society*? It appears that in the course of conscious evolution this fundamental evolutional legacy will be absorbed in the mentality of individuals who make up a mature society, and will become the property of all mankind. In other words the cultural space in mature society will be formed following the synergetic principle of integrity and individuality of culture phenomenon. Under these conditions its integrity will be determined by the values common to all mankind and individuality by mental and intellectual peculiarities of the individual.

Thus the *mature society* in essence will be a "factory" of individuals and a guarantor of sustainable future. Each group of people will strive to reach it through their mentality and culture, that is national values and the values common to all mankind. In mature society a human factor will become dominant in the existence and conscious evolution of Life rather than technics and technology.

Thinking is a self-organizing activity that is a qualitatively new level of human brain functioning, which resulted in the beginning of the cognitive stage of Man's intellectual evolution. *Thinking* as a phenomenon of self-organization phenomenon is the highest hierarchical level of consciousness and gives it the higher extent of intentionality. At the same time the phenomenon of brain self-organization has opened "a door" to unordinary states of consciousness which manifest themselves in such psychological anomalies as splitting, para-psychological phenomena, virtuality, etc.

A thought which may be both real and abstract is the product of thinking. It is through thinking that a Man became capable to apperception that is a conscious perception of reality and integral world outlook through spirituality. *Thinking,* as a conclusion, may be of two kinds: a deduction by analogy (animatism, archaic *thinking,* comparative analysis) and logical conclusion (analytic *thinking*). In normal state the archaic *thinking* through spirituality, taking root in Self, creates the integral world of a Man, while analytic *thinking,* abstracting the real world expands and deepens knowledge of objective reality. The disturbance of their synergism produces alienation which is manifested in such anomalies as the splitting of the psyche, fanatism and virtuality.

In terms of morpho-physiology *thinking* is caused by the phenomenon of human brain independence that is by its self-organizing activity. It is possible it can also explain a phenomenon of brain cerebralization in a contemporary Man.

In the future, the brain of a self-sufficient Man owing to his evolution will become the sixth receptive centre, that is, it will function in the creative, constructive regime. And this will overcome the problems which appear now because of the ever-increasing rate of changes to conditions of Man's existence (*being*).

Scientific religion (synergetic world outlook, new world view) – faith in scientific knowledge that nature in a broad sense (the universe) – is the self-organizing chaos "with no beginning or end". Synergy of living nature and a Man is their fundamental quality. And Life is a unique form of self-organization, evolution of which has led to self-insight and principally new world outlook. *Scientific religion,* as a qualitatively new kind of spirituality can arise only through the scientific understanding of Life's origin and its place in the universe.

Thus, the basis of *scientific religion* is the synergetic world outlook. A Man, having a synergetic nature, proves to be integral not only in the world of living nature, but in the universe in general. That is why the scientific religion in essence is the faith in a new scientific picture of the world, according to which the universe is open, synergetic and integral. In contrast to the mechanistic and physical models of the world this new synergetic picture, like essence, includes the phenomenon of consciousness★ in the form of a phenomenon of self-organization. The consciousness becomes a fundamental property of the universe and thus completely overcomes the limitedness of physical and mechanistic models of the world. That is why, in our opinion the Big Bang theory and theory of an expanding universe is just another scientific abstraction and illusion, since it comes from the erroneous image of the world (built of elementary particles – bricks of matters), as well as of archaic religious and philosophic ideas on the existence of the original cause, and illusions on transcendence of the causation principle, accepted in contemporary science. This paradigm is particularly materialistic, and there is no place for the phenomenon of self-organization in it. And so, the necessity for its complete replacement becomes more and more evident.

In the synergetic picture of the universe, according to which the world is the self-organizing chaos "with no beginning or end", the existence of conception causes is completely negated. Besides, the synergetic world is not a determined but probabilistic world. Thus it is natural that a synergetic approach is a qualitatively new level of world outlook which discovers a Man's new treatment of the world and of himself. As a result he becomes self-sufficient, that is an individual, and by religious canons – the enlightened Man.

So, *scientific religion* is one without dogmas, without divine origin, without hell or heaven, with no holiness, providence and worship, i.e., it is not based on mysticism, blind worship, but on the knowledge of the scientific picture of the world, and is Man's goodwill and honouring of Life and his responsibility towards its continuous perfection.

Scientific religion is a new spiritual epoch, a higher level of world outlook being characteristic of it. By removing the existing illusory idea of antagonism between science and religion, it synthesizes them i.e., makes a great synthesis of morals, humanism, science and scientific-technical progress.

And, at last, the *scientific religion* which synthesizes within itself spirituality and science, that is faith and knowledge, opens principally new horizons of humankind's development aimed at archetypical ideal, absolute perfection, the Absolute – the conscious step of Life evolution. In other words *the scientific religion* is the world outlook of a self-sufficient Man, and the Absolute in it is the perfect individual.

Thus the *scientific religion* is neither religious, in a classical sense of this word, nor metaphysical or positivistic, but a qualitatively new, synergetic world outlook. That is, according to *scientific religion,* nature has a synergetic basis. It should be noted that in contrast to traditional scientific notions, the synergetic one is based on the principle – the whole is always more than a sum of its parts. Thus the realization of the world integrity by a Man means the adoption of a scientific religion and synergetic world outlook by him. So, it is thanks to scientific religion that a contemporary Man will be capable to conscious understanding of his place in the Universe. And thus a new hierarchic level of adaptation will be revealed, that is the conscious adaptation of a Man.

Humanized education is the education system based on a synergetic view of the world. The conscious evolution and natural need for permanent perfection of Man's *being* will be its core. This education equally negates the theological, materialistic, idealistic, positivists' views of the world. None of these outlooks can be the full basis of a world outlook for it, since the real world of a contemporary Man is far beyond their limits. Besides, the humanized education, in contrast to the existing systems will be directed to the future rather than to the past and present and it will serve as the basis for perfection of the human *being*. Students will take an active part in the process of learning by dialogue and cooperative creative activity. And accumulated experience and knowledge will only be the reference information. Thus, educational institutions in future will be generally of an enlightening character and aim.

The ideal of *humanized education* will be absolute perfection, the Absolute, which opens the way to self-insight and the formation of self-sufficiency, that is Man's formation as an individual. In other words, in the system of *humanized education* the students will learn to create their future. So, its ultimate aim, in contrast to the present education systems, is not the training of a positivist, a characteristic of today's education system, but of a self-sufficient man (individual) with the capability to foresee the future. Figuratively speaking, the system of *humanized education* will become a "factory" of individuals where the synergetic view of the world, i.e. scientific religion will be preached, and students will develop and perfect their creative activity towards absolute perfection, the Absolute.

Perception and apperception. Perception is a qualitatively new hierarchical level of understanding reality, that is, its perception through sensory images (or types of senses characteristic only to a Man). Being a result of synergism of receptive senses it has determined the origin of *Homo perceptions* (anthropithecus) and bifurcation of biologic environment. Thus, a principally new stage of evolution was onset: intellectual evolution of Life.

The original *perception* was sensory images of biological instincts. Then, in the process of evolution of the anthropithecus these images were formed as relic archetypes and were fixed in his genetic memory as a universal human attribute. In analogy with instincts themselves the relic archetypes attach to the anthropithecus' actions the intentional (directed, active, labor) character. And what is more, in the course of intellectual evolution these relic archetypes, being sublimated into the matrix of archetypical memory of *Homo habilis*, have opened a new phase of Man's evolution.

In terms of morpho-physiology, the *perception* is a result of the origin of a new structure of brain cortex containing relic archetypes. Meanwhile, if *perception* is uncomprehended perception, i.e. sensory cognition, then apperception, which arose in the following stages of Man's evolution, is a comprehended perception – rational cognition, since, in addition, it includes the identification or revealing of an analogy of perceptive images with images, signs, symbols and notions already existing in the archetypical memory of a Man. Thus, if perception produced the phenomenon of "sensing-in" (understanding) then apperception is a comprehended understanding, a conclusion by analogy.

Populational (or morpho-physiological) adaptability – is adaptability of Life, additionally including a principally new and highly efficient method of adaptation to the environment by reproductive modification of the genotype of population's individuals. This adaptability is based on the fact that living organisms possess a sufficiently high genetic evolutional potency, that is they possess high evolutional flexibility for the reproduction of progeny capable of optimal adaptation to the changing conditions of the environment. Meanwhile, it is not mutation, but sexual reproduction, and, respectively, an ability of reproductive isolation lead to the origin of a new biologic species. Thus, it is sexual reproduction which opens unique opportunities for the population or morpho-physiological method of adaptation of living organisms to changing conditions of the environment.

Psyche – is a reflection, in a broad sense of the word, of the phenomenon of intentionality of consciousness of living creatures on the objects of inner and outer reality. In a narrow sense, i.e., on the level of a contemporary Man, *psyche* is also a perceptive, apperceptive and thinking activity of his brain. A specific quality of the contemporary Man's *psyche* is its ability to seperate into the consciousness and the unconscious (archetypical). This phenomenon is accompanied by spiritual crisis, having both positive and negative aspects. A positive aspect leads to spiritual growth and negative one – to mental and psychosomatic disorders.

Instincts are the primary basis of *psyche*. The contemporary Man's psyche usually has two phases in ontogeny: childish (immature, archetypical) and mature (spiritual). In the future, according to the synergetic theory of Life, the third phase will form in a self-sufficient Man – the enlightened *psyche*, i.e., conscious *psyche* deprived of its unconscious component. It will be free from separation or other analogous anomalies, inherent in *psyche* of a contemporary Man. Thus, the spiritual crisis and psyche separability are fundamental peculiarities of *psyche* of a contemporary Man. One should also note the psycho-forming role of children psychology, being of archetypical nature.

In terms of morpho-physiology *psyche*, as the result of intentional activity of consciousness, is the method of *being* of living creatures. And archetypical phenomena (relic archetypes, Self and archetypical ideal – absolute perfection, the Absolute) serve as the bioenergetic channel of psychology of a contemporary

Self-sufficient Man (superanthropus, individual) – a new species of a Man, which will originate through self-insight and acceptance of synergetic world outlook which arouses a qualitatively new type of spirituality – scientific religion. In phylogenesis he will be a superanthropus. He will feel inside and will be sure in the integrity of his existence in the universe, and he will proceed from this notion as an attitude and a moral norm common to all mankind.

Besides, the phenomenon of integrity in his world outlook will include not only the world of living nature but also the universe in general. In other words, spirituality based on the blind faith as well as mythological, mystical, religious and other forms of world outlook and, respectively, pictures of Man's existence integrity created on its basis lose their power for a self-sufficient Man.

At the same time as a result of self-insight and adoption of scientific religion he will acquire his integrity in the universe, his spirituality at qualitatively new level of world outlook, that is, he will not understand his integrity in the universe through myths, blind faith and religious values but through knowledge of the scientific image of the universe according to which the world is an open self-organizing chaos. And a Man will see himself as indivisible part of the universe. Such rebirth of a contemporary man into *self-sufficient Man* will lead to conscious bifurcation of mankind, since the system of values of a *self-sufficient Man* will radically differ from that of a contemporary Man. The role of religious spirituality in it will already belong to archetypical ideal – absolute perfection, the Absolute.

Thus through self-insight and scientific religion a contemporary Man will not only comprehend himself, the purpose and meaning of Life but will consciously understand his place, his status in the world. As a result the synergetic world outlook and the absolutely new kind of adaptation, that is the conscious adaptation will form in him. Besides, his intentionality of his consciousness will become conscious, that is it will be determined by archetypical ideal – absolute perfection, the Absolute. A Man will become self-sufficient, that is himself, who he really is according to his nature. Breaking away from any idols and idolatry, in particular his worship of the wealth of scientific–technical progress, he will become an individual in a full sense of the word. And thus the Life evolution will, at last open its purpose and meaning – to be self-sufficient and to strive to absolute perfection, the Absolute. And a Man will understand his predestination – to preserve and perfect Life. The self-sufficient Man's *being* and way of life will be based on the canons of Life evolution, but the main thing is that a Man himself will become a driving force of its new stage – the conscious evolution of Life.

That is why the care for mankind will be more important for him than interests of separate individuals, families, political parties and nations. Obviously, this will radically change Man's relation to himself and open new opportunities in his constant self-perfection. The existing ambiguity in understanding a phenomenon of individuality will be completely removed. It becomes clear, according to STLife, that each person is potentially an individual, and the way of the individual's formation is universal: self-insight, adoption of scientific religion and synergetic world outlook.

It is customary to think that the model of integrity (spirituality in a broad sense) perceived by an individual is his fundamental quality. But this integrity and spirituality may be of absolutely different origin and scale: religious, moral, traditionalist, national, culturologic, etc. This exactly is the reason why the "partial" Man is always present in the available notions about an individual. But in contrast to these "partial" individuals *a self-sufficient Man* will be integral in essence rather than in separate aspects of his *being*, and as a result he can be an individual in the full sense of the word. Hence, *a self-sufficient Man* is a spiritual, enlightened Man, who has overcome himself and has become a symbol of integrity perceived by him that is the synergetic picture of the universe. In so doing the spirituality, that is a phenomenon of *self-sufficient Man's* overcoming of himself, will be based not on the blind faith but on knowledge of the scientific picture of the world. And Man's evolution will already be conscious.

Thus, an individual in a full sense of the word is *a self-sufficient Man*, who has undergone psycho-synthesis of partial individuals, has overcome himself and has found his integrity in the universe. His archetypical ideal will be absolute perfection, the Absolute. That is why the existence of *a self-sufficient Man* is directly connected with his creative activity, that is with his striving to this absolutely new form of archetypical phenomenon. *A self-sufficient Man* is a creative creature, possesses new numerous qualities, absent in a contemporary

C. Young wrote: "My life is permeated by the only idea and is concentrated on the only aim, that is: on penetration into the individual's mystery. Everything may be explained from this central point, and all my work is connected with this subject matter". Unfortunately, he had not succeeded in finding the individual's mystery, and his scientific strivings have not reached the objective, since his conception of individuation – a process of individual's formation and maturation by means of assimilation, "capture" by his consciousness the content of personal and collective unconscious is, in our opinion, profoundly erroneous. Since the phenomenon of individuation does not stimulate the process of assimilation, but vice versa, it even deepens the phenomenon of mental splitting of the contemporary man into consciousness and unconscious. As to the process of assimilation, it really proceeds not by individuation, but through its spiritual component, by leap-like development of intellect, that is spiritual renewal.

A self-sufficient Man as an individual will consciously be the master of his own destiny. First of all he will renounce the priority of the world of values created by him in favor of the really existing world. Life evolution will find in his person its purpose and meaning, expressed in his striving to absolute perfection, the Absolute. That is why he, in contrast to a contemporary Man, will search for his present, purpose and meaning of his existence in the future rather than in the past. His adaptation will be conscious that is his *being* will be oriented to the ideal, absolute perfection, the Absolute, and this will define his essence.

Naturally, the self-insight, acceptance of scientific religion and synergetic world outlook will be reflected in other specific peculiarities of *a self-sufficient Man* as well. In particular, the universe in his world outlook will be not only open but also synergetic, and a Man will only be an indivisible part of this integral world. Moreover, the subject-object division of the world, widely used in the modern science and philosophy, will lose any meaning for him and will become a fiction. In reality the universe will be open, integral and synergetic for *a self-sufficient Man*. This new scientific picture of the world radically differs from the existing philosophic world outlooks. One should think it is rather natural, since the latter are not directly connected with reality, but were formed by the abstracting property of the phenomenon of thinking, that is they are based on abstract notions, categories, conceptions, which have in principle exhausted themselves long ago. In particular, the ideas of a Man as the aggregate of a body and spirit, *being* and consciousness, etc., used by philosophers are now rather naïve, and deeply erroneous, since firstly, the unconscious part of psyche is absent in these "formulas", and secondly, the contemporary Man, as to his nature, is the integral, indivisible spiritual creature. That is why neither idealism nor materialism, ontologism, phenomenologism, existentialism or other analogous philosophic teachings can no longer serve as the world outlook for him.

The achieved level of human intellect is so high that philosophic abstractions, forming its basis, have become illusory. If materialism recognizes that the matter, being a certain abstraction, is primary and consciousness is secondary, idealism, on the contrary, considers that the absolute spirit, absolute idea is primary and material world – secondary. The contemporary philosophy, in contrast, confirms the primacy of *being*, and its trends such as positivism, phenomenologism and pragmatism totally negate the cognitive and ideological value of philosophic investigations.

When discussing the limitation and immaturity of philosophic teachings the contemporary wits say: "there are as many philosophic views in the world as people". But in essence, the history of philosophy is the history of a search for new spirituality. So, a critical situation which has arisen in philosophic investigations can be overcome only with a new level of abstracting with incomparably high intellectual insight and generalizing (integrating) quality that is due to leap-like development of intellect – the spiritual renewal.

The synergetic world is sufficiently complicated since it is non-equilibrium, unstable and irreversible. Besides, it possesses a capacity towards dynamic equilibrium. In particular, symbiosis of living creatures is widely spread in the world of living nature. Thus, the multi-essence and ambiguity are peculiar to the synergetic world i.e. multi-variance of the way of its development.

Naturally, the synergetic world outlook will be radically renewing the way of understanding of both outer and inner world by a *self-sufficient Man*. Since, owing to knowledge of the scientific picture of the world, Man's intuition as the merging of a subject with an object will be a certain analogue of receptive senses and will be of mental character, *a self-sufficient Man*, in contrast to a contemporary Man, will be able to accept the universe as a whole and consider it as his habitat. That is why his perceptions and intellect will be called enlightened, since the enlightened Man's brain becomes qualitatively new, the sixth receptive centre. This greatest result of intellectual evolution of Life could further serve as a fundamental basis of Man's conscious evolution. Under these conditions the orientation and *being* of a *self-sufficient Man* will be absolutely different than the orientation and *being* of a contemporary Man.

Another fundamental property of a *self-sufficient Man* will be his archetypical ideal, absolute perfection, the Absolute as a sublimation of the archetypical phenomenon of Self, being the basis of religious spirituality. Such an ideal is not a result of rational reasoning, but the archetypical ideal, that is the ideal directly connected with Life evolution. That is why *a self-sufficient Man's being*, in contrast to that of a contemporary Man, will be creative, constructive and directed to perfection, the Absolute. Of course, this evolutional metamorphosis will be radically changing the psychology of a *self-sufficient Man,* since just this ideal will become its centre, nucleus, rather than the previous archetypical phenomenon of Self, being a symbol of understanding the existence integrity of a contemporary Man in the world.

Psyche of such an absolutely new kind can be called the enlightened psychology; it will be based on knowledge of the scientific picture of the universe by a *self-sufficient Man* and clarification of his place in it. And this means that the enlightened psychology will not be latent and unyielding to rational cognition, but will become a reality and indivisible, since it will be based on conscious world outlook, being more developed and perfected than existing integral and of value world outlook of the reality, which is based, respectively, on the relic archetypes and blind faith. In this connection the enlightened psyche of *a self-sufficient Man* will be integrated, that is its conscious and unconscious aspects will be merged, integral. So respectively, there will be no place in it for mysticism, myths and other forms of irrationality.

Evidently, the enlightened psychology will also be radically changing the way of adaptation of a *self-sufficient Man,* that is: in contrast to a contemporary Man, the spiritual way of adaptation being peculiar to him, he will adapt consciously on the basis of knowledge of the scientific picture of the world, and intentionality of his consciousness will be expressed in striving to absolute perfection, the Absolute. Meanwhile, the Life evolution in the person of a *self-sufficient Man* will first find its purpose and meaning – absolute perfection, the Absolute. In this connection he will consider the birth of children not only as the biologic act of continuing an individual's life, but also as vital necessity of his continuous perfection and as the method of solving the eternal problem – the problem of immortality. *A self-sufficient Man* will be aware that after biologic death he will continue living in spiritual, cultural principles of progeny as the basic principle for further self-perfection, as well as for bringing-up progeny more perfect than himself; and that is Man's immortality. Thus, the striving for the Absolute and eternity without any mysticism, God or the equivalent highest value becomes the essence and natural need of *a self-sufficient*

Such social and psychological anomalies as the lack of spirituality, mental splitting, inner fear of the future (shock of the future), alienation, leaving for the virtual world, collapse of adaptability and other analogous negatives inherent in the present crisis of a contemporary Man will be alien to *a self-sufficient Man* with these principally new qualities. A completely conscious world, the absolute perfection, the Absolute being ideal in it, rather than an archetypical world, the world of unconscious, will be present in the psyche of *a self-sufficient Man*, in contrast to that of a contemporary Man. What is more *a self-sufficient Man* will be able to create himself as a perfect individual★ and ensure his conscious spiritual immortality through the humanized education. Thus, he may also be called an enlightened From the viewpoint of ethics, *a self-sufficient Man* will never worship anybody and anything. He can only bow out of respect, since he will live on the basis of knowledge, rather than on blind faith.

Thus, the brain of a *self-sufficient Man* will become a qualitatively new, sixth receptive organ, and owing to the subject–object syncreticity (merging) the intuition (direct judgment) will play the main role in his cognitive activity. And he, as an enlightened individual, will break free of all sins of psyche of a contemporary Man and will strive to absolute perfection, the Absolute.

Independence of human brain – a qualitatively new hierarchic level of functioning of human brain which has opened his self-organizing activity, that is, thinking. This phenomenon is a result of synergism of relic archetypes and origin of a contemporary Man as a new intellectual species, neoanthropus. His unique quality is the archetypical phenomenon of Self, symbolizing the understanding of the world integrity by

Self in the contemporary Man's consciousness manifests as spirituality. And if all other living creatures are dissolved ecstatically in their environment, that is they are merged with the habitat, the integral relation with the surrounding is created in a contemporary Man owing to this spirituality. In this connection he, on the one hand, becomes isolated in respect of the surrounding world, and, on the other hand – due to apperceptive ability a Man becomes both a subject understanding his integrity and a cognizing subject. That is why he obtained a possibility of sensory use of totems – animals, plants, objects and natural phenomena as spiritual objects of worship.

Thus, evolution has found a qualitatively new way of adaptation, that is spiritual way of adaptation to the surrounding world. That means that at this stage of intellectual evolution of Life the morpho-physiological (genetic) mechanism of evolution has exhausted itself and a Man began to evolve on spiritual basis, that is spirituality became a necessary condition of adaptation of a contemporary Man in the surrounding world. This became the start of the new, cognitive stage of intellectual evolution of Life.

It should be noted that totem spirituality in the further process of intellectual evolution was transformed into religious spirituality. Something analogous occurs even in our time, when religious spirituality, as a result of psychological phenomenon of Self, loses the attractive force and ersatz spiritualities are created through neoidolatry (worship of wealth of the scientific and technical progress, money, idols, etc.), a qualitatively new spirituality will be created within the above spiritualities. According to STLife exactly that will be the scientific religion based on knowledge of the scientific picture of the world.

Thus, a contemporary Man is not virtually *Homo sapiens* (intellectual Man), but a spiritual creature. Thus, under the lack of spirituality, which is characteristic of the present, he will certainly proceed towards folly and decadence. Created by Man, atomic, hydrogen and neutron bombs as a means of mass destruction of people and their habitat can serve as symbols of folly and unreasonableness of the contemporary Man.

It should be specially noted that Man's perception and connected with it alienation from the surrounding reality, being his distinction as compared to animals create his specific world that is the world of values: biologic, spiritual and intellectual. Owing to the fact that spirituality* is the absolute value of existence of a contemporary Man, this world is arbitrarily structured as the individual system of values which, when knit with biologic values, becomes a factor determining the *being* and behavior of a contemporary Man, neoanthropus. That means, the behavior of a Man normally is not determined by intellect, as is customary, but by archetypes and individual system of values. That is why the human inclination is not of the same order as that of animals, since they, in contrast to a contemporary Man, have no spirituality and intellectual world of values. In short, if inclination of animals is of instinctive nature, the human inclination is of the value one.

Spirituality as archetypical phenomenon is inherent in a contemporary Man only. Awaking images, symbols of existence in his consciousness, it carries him away, seizes him and thus forms a qualitatively new way of behaviour. It is more complex and multifactor, therefore a simplified scheme "stimulus-reaction" taken in behaviorism is absolutely unacceptable in a case of human behavior. But the fact that the *being* and behavior of a contemporary Man are not determined by intellect, but by value inclination, to say more exactly, by individual system of values, that allows one to reach a conclusion that he still remains immature and imperfect in his nature. In this connection it should be emphasized that the notion *Homo sapiens* – an intellectual Man, introduced by old-Greek philosophers and attributed to a contemporary Man, is the greatest delusion and has catastrophic consequences for all mankind, since exactly it serves as the basis of nihilism, narcissism, unrestrained money-grubbing of a contemporary Man and his striving to be a master of nature.

Immaturity of a contemporary Man and vulnerability of his psyche manifest themselves rather clearly in our time of spiritual crisis. The lack of spirituality is first of all the loss of understanding of world integrity by a Man, and   leads to his adaptational collapse, instability and mental splitting. And all that is the cause of such anomalies, as alienation, decadence, leaving for the virtual world, degradation of morals, traditions and family, as well as other social and psychological anomalies. In a word, a contemporary Man under the lack of spirituality, losing the understanding of the world integrity, becomes in the end a Man-Narcissus. Being alienated from reality he finds himself in the dead alley of evolution. He does not know that yet, but feels rather sharply.

To avoid the factors threatening safety of his existence, a Man feverishly looks for a new form of spirituality or tries to substitute spirituality with ersatzes (worship of science, sovereignty, utopian ideas of more perfect social orders, etc.), which produce most different social myths and neoidolatry in particular, worship of wealth of scientific and technical progress. It is interesting to note that in this connection there exists a certain similarity between activity of archaic mages and contemporary scientists. If the former created the totem idols, which engendered paganism, the contemporary scientists create technical and technological idols and thus revive idolatry of new quality, that is neo-idolatry. But in both cases that means the crisis of faith, crisis of spirituality. To overcome such crisis situation and provide the integral world outlook required for adaptation, more refined types of abstractions are widely used in both cases. But the level of abstraction in the archaic magic is much more primitive than the abstraction level in modern science.

As is known, spirituality is an absolute value for a contemporary Man, its loss may be replenished only by a new more developed, more perfect form of spirituality. And the fact that today spirituality on a mass scale is substituted by neo-idols, commodity goods in particular, that is the material wealth, and a contemporary Man degenerates and turns into materialized Man, and this is not just decadence but a presage of the upcoming in the near future catastrophe, Life deadlock in general. Thus it is not excluded that a contemporary Man, when losing his will and locking-in himself can very much become a cause of the mortal disease for Life on the Earth.

It should be also noted that a materialized or one-dimensional Man remotely resembles Skinner's "rat-morphic" Man whose behavior is like the rat's behavior. In correspondence with this a possibility of psychic manipulation by people is presumed by externally imposing new needs on them. And thus, an illusory idea of the society of abundance and general prosperity is created as the way of achieving happiness and calm. That is exceptionally dangerous delusion, since it turns the economic growth into the end in itself and source of calamity.

Syncretic science is a qualitatively new hierarchic level of science based on a synergetic view on the world characteristic of a self-sufficient Man★. As a result of conscious bifurcation and vital emotional experience, by him, a new image of the universe will be formed in a Man as a predisposition and as an aim on a new image; as a consequence he will rise to a qualitatively new level of world outlook. The world will become open and synergetic for him.

This image of the world is far beyond the limits of today's objectivistic science and persuades us of its extreme limitation. In particular the available pictures (mechanistic, physical, biologic, etc.) of the world, created by objectivistic science, are based on the matter-energy ground and there is no place for the phenomenon of consciousness in them. And in a new synergetic picture this principally important problem is solved automatically, since the phenomenon of consciousness within the new scientific picture of the world becomes all-embracing natural property of the universe. Under these conditions the limitation of modern objectivistic science in the integral understanding of nature and a Man is completely taken off, opening the way to self-insight and qualitatively new level of *being*. And the main question of philosophy – relation between matter and spirit, body and consciousness – lose any meaning.

It should be also noted that the modern objectivistic science as to its nature is based on the subject method of reality representation, that is on atomists thinking, search for fundamental principles, initial cause. That is the great delusion, since the universe in reality is open, that is the self-organizing chaos "with no beginning or end". Also the notion of time has sense only in self-organizing systems, where the order prevails over the chaos. The order and chaos, otherwise the birth and death exist simultaneously in the universe and they are inseparable. If the universe is somewhere ordered with coming of self-organization, then its order breaks up in some other place with the advent of chaos. Thus the crisis of modern science becomes more and more obvious; it is expressed in the restriction of its world outlook and mystifiedness.

The discovery of the countless number of elementary particles, finding out of greater quantity of new exotic genes, such as a gene of adultery, gene of organism death as well as the projects of utopian social orders and political ideas, e.g., the society of sustainable development, the society of one-fifth, the society of risk, the idea of collision of civilizations (cultures), as well as other analogous "scientific" products of illusory mystic character – are not just innocent delusions. There is a real danger of assimilation, capture of some of them by public consciousness that can prove to be a catastrophe. In particular, in conditions of globalization of human life activity the vicious scientific idea of collision of civilizations (cultures) acquires a real ground, and a sociologist cannot offer a reasonable alternative. In this connection ideological struggle or cold-war policy has passed from the social sphere to spiritual. That is why that in conditions of the deepening lack of spirituality it is extremely necessary to put on one's guard against mystifications of the modern science.

The idea of "objectivity", which has been led by modern science to the absolute, will be abolished once and for all, since the synergetic world outlook completely devaluates the objective-subjective separation of the world and negates transcendence of space, time, causality, determinancy and other analogous artificially created intellectual constructions, categories, etc. Nevertheless, the abolition in this case means that the canons of objectivistic science will lose their former illusory universality and absoluteness, become norms, rules for separate partial cases and will be considered as a certain approximation. This, first of all, will apply to the study of physical world. Simply speaking, it is the synergetic picture of the universe which will become a "legislator" in understanding of the universe, natural and social phenomena, rather than the principles and laws of physics or any other field of natural-technical sciences, since precisely it (the synergetic picture or the universe) makes the basis of everything existing in the real world.

Thus, the *syncretic science* will be rather oriented to synergetic world outlook than to materialism, idealism and positivism.

It is clear from the above statement that a cardinal change in relation of *a self-sufficient Man* to the surrounding world and himself, multi-variant character of development of reality and multi-essence of *being*, that is the synergetic world outlook will be of pluralistic and probabilistic character. Besides, the synergetic picture will awake qualitatively new and more developed methods of perception and thinking in *a self-sufficient* As was noticed above, they will be enlightened perception and metathinking (or algorithmic thinking) in him, and this will open a qualitatively new level of cognition. New science will appear which can be called *syncretic science*.

In contrast to today's objectivistic science, which eliminates all other ways of cognition, besides the scientific one, the *syncretic* science, on the contrary will integrate scientific knowledge with art, literature, philosophy and other humanitarian ways of world outlook as essentially important additional possibilities in studying and recreation of the most complete and adequate images, models and pictures of reality. One should think that only due to such a syncretic approach, will it be possible to form a really all-embracing scientific notion and to open a way to constant perfection of a Man and quality of his life.

Art serves the understanding of the world, and the humanities – the understanding of Man's *being*. Thus relationship of the objectivistic science to the art and humanities is equivalent of relationship of knowledge to understanding. And knowledge and understanding are two absolutely different categories. The understanding develops from the general to the particular, while knowledge on the contrary – from the particular to the general. What is more, if knowledge is based on science, the understanding – on the Man's world of values.

People understand easily each other, if their individual system of values is close. According to its nature, the adaptation of a contemporary Man is directly connected with understanding, religious spirituality rather than with cognition and knowledge. However a Man, who has erroneously exchanged the spiritual way of adaptation for the adaptation based on his domination over nature that is the negative adaptation, has lost the evolutional way of development. This mechanistic substitution of understanding with knowledge has caused the transformation of a contemporary Man into the destructive element of Life. And the fact that the objectivistic science cannot integrate with art, religion is a result of ideological degradation of a contemporary Man. This is illustrated especially obviously by interrelations of objectivistic science and religion. A conflict always existed between them, but it is rather nominal and meaningless, since it is a result of the use of two different, mutually exclusive conceptions of a Man by them.

Today, a contemporary Man, when transforming into the positivist, completely loses his ideological reference point and locks-in on himself.

The syncretic science, being based on synergetic world outlook, substituting today's objectivistic science, will automatically remove its limitations (the above ones and similar to them) and will serve as a principally new way of adaptation of a self-sufficient Man in the conditions of conscious evolution. The driving force of this new form of evolution will be the internal striving of a self-sufficient Man to archetypical ideal – absolute perfection, the Absolute. And cognitive process in the *syncretic science* will be based on synergetic world outlook, intuition and scientific-technical and technological creativity. Meanwhile the pragmatic aspects of scientific problems, that is the development of algorithms of Man's activities on the improvement of life quality, will become the priority objective rather than the fundamental-theoretical investigations directed to world outlook. In other words, the *syncretic science* will be rather of a constructive and active (that is, of pragmatic character) than of the scientific-cognitive one.

Knowledge obtained from the syncretic science will serve as the algorithmic element of *being* of a self-sufficient Thus, a perfection of Man's *being* rather than world outlook will become the main function of the syncretic science. In other words, the scientific activity will not be an end in itself (that is, science for the sake of science) but will be a necessary condition of perfecting Man's *being* and growth of individual.

Further intellectual development of a self-sufficient Man will no longer destroy spirituality (that is, the scientific religion) but in contrast it will deepen and perfect it. This means the simultaneous deepening of the scientific religion and creation of conditions for Man's development without crises, i.e. Life evolution will become conscious. The *syncretic science* will become its immaculate instrument.

Consciousness is a phenomenon of self-organization in a broad sense, and in the narrow sense it is Life, being its unique form. Its highest hierarchic level is thinking, being a self-organizing activity of the human brain. According to STLife, hierarchy is the basic principle of self-organization phenomenon:

physico-chemical kinds of self-organization $\longrightarrow$

$\longrightarrow$ biological kinds of self-organization $\longrightarrow$

$\longrightarrow$ self-organization of the human brain,

that is, a more perfected kind of self-organization arises on the basis of less perfected kinds.

The world is open and synergetic at the level of a self-sufficient Man, and *consciousness* for him is the initial fundamental property of the universe. That is why the so-called main question of philosophy is meaningless for him, since the matter and spirit, *being* and consciousness are inseparable attributes of the universe. It should be also noted that the notions "soul", "psyche", "consciousness", "unconscious" still don't yet have an unambiguous interpretation. In particular, in modern science psyche and consciousness, psyche and soul are considered rather equivalent and *unconscious* is excluded at all. According to STLife, a soul – is unsplit psyche, and psyche itself is a degraded soul or that containing evil spirit which consists of a rational consciousness (thinking) and unconscious. That is why psyche of a contemporary Man is far from a complete synthesis, and this is the main cause of his imperfection. Evolution of consciousness, similar to intellectual evolution, is accompanied by non-ordinary states of *consciousness*, that is by states of spiritual crisis and spiritual renewal, which is erroneously perceived by modern science and medicine as the mental disorder. And the lack of spirituality, loss of soul and psyche split of a contemporary Man make the basis of the appeared deadlock of Man's evolution. The main thing is that a principally new notion of consciousness, proposed by us, completely invalidates all existing religious, philosophic, scientific, psychological, socio-economic and other analogous ideas of a Man.

Biologic values and value adaptation. Biologic values are relic archetypes being perceptive images (perceptions) of biologic instincts. They awake the kinds of senses inherent in a Man only. Giving the human consciousness the attribute of intentionality (direction of consciousness to the object) they turn Man's action into activity, into labor. In terms of psychology the biologic values are expressed in a form of images and symbols. And the main thing is that they serve as the basis for qualitatively new way of adaptation that is the value adaptation of a Man to the surrounding reality. In the course of Man's evolution on the basis of relic archetypes there has arisen an archetypical or bioenergetic channel of psyche of a contemporary Man:

relic archetypes ➤ Self ➤ spirituality − − ➤

➤ archetypical ideal, absolute perfection, the Absolute.

Here the dotted arrow depicts the future. Thus, a Man, in contrast to other living creatures, has archetypical (psycho-physiological) mechanism of evolution. And this mechanism is more developed, more perfect, than the metabolic (cellular) and genetic (biologic) mechanism of Life evolution which appeared before.

A contemporary Man*; positivist – a spiritual creature, whose *being* and behavior are determined by the individual system of values, which take roots in relic archetypes, that is in instinctive correlates. From the viewpoint of evolution it has been existing as a new biologic species for about 40,000 years. *A contemporary Man* in essence is not *Homo sapience* (intelligent Man), but a spiritual creature, and a spiritual way of adaptation determines his essence. In this connection the spiritual crisis of *a contemporary Man* is his evolutional crisis, and in normal state he will not be able to live without spirituality.

This fundamental feature of *a contemporary Man* is closely related to his mystical and myth-creating activities and appearance of great religions. Myths, being a product of conscious understanding of reality by a Man, serve an effective method of overcoming the deficit of knowledge during the formation of integral world outlook. And the integral world outlook, being awakened by spirituality, and respectively the creation of order from chaos is, as is known, a necessary condition of adaptation of *a contemporary Man* in the surrounding world. So, it is no coincidence that all existing religions are based on myths and religious cults, which have special value in creation of the image of integrity of human existence in the world.

According to the synergetic theory of Life two hierarchic levels of self-organization – biological and archetypical, function simultaneously on the organismic level of *a contemporary Man*. In morpho-physiological context the vegetative nervous system providing somatic integrity and highest nervous system responsible for brain self-organization, that is thinking, correspond to them. Their integrity (syncreticity) is based on archetypical phenomenon of Self, which is the basis of the spirituality of *a contemporary Man*. A Man has discovered his duality in most ancient times, when he understood that a soul lives in his body, independent of it. Further on a Man considered himself as the complex of a body and a soul. That was the greatest discovery, which had considerably increased the rate of Man's evolution. That is why in terms of evolution *a contemporary Man* is far from maturity, since he does not know his own nature, as a result his world view is also ambiguous. And if the biologic basis of *a contemporary Man*, except for such evolutionary rudiments (relics) as color of skin, figure and other anthropological features, is uniform and common to all mankind, his world of values and spiritual principles are diverse. That is expressed in language, faith, culture, mentality, and other spheres of Man's social life.

It is interesting to note that the independence of Man's *being* from the canons of logic and laws of modern objectivistic science, i.e. his existentiality, is explained also by duality of *a contemporary Man*. In this connection, the empirical statement that the *being* determines the consciousness has no scientific basis, since the notion of "consciousness" here is understood in the Cartesian sense, i.e., it is considered to be completely equivalent to Man's psyche. That means that a given formula does not take into account an unconscious component of psyche, which in many aspects determines the *being* and the behavior of *a contemporary Man*.

In phylogenesis a contemporary Man is neoanthropus, the apperception and conscious world outlook being inherent in him. That is why in conditions of degradation of religious spirituality, which is occurring right now, the spiritual symbols, being the absolute values and system-forming factors of the world of value of a *contemporary Man,* are more and more substituted by ersatz symbols: scientific-technical, social, economical, political and ideological ones. But this substitution does not solve the problem, but drives it inside, accelerating the process of spiritual degradation of *a contemporary Man*.

Consequently, a Man constantly loses such qualities as morality and spiritual way of adaptation. The paradox is that though he knows the reality, the external world is incomparably bigger and deeper than the one his religious ancestors of the Middle Ages imagined. But in contrast to them because of the loss of spirituality, a Man does not understand the integrity of the world, and the main thing is that he does not know his place in it. A Man of our time has become even more problematic, since he does not understand the world and thus cannot adapt in it. In other words, the phenomenon of "understanding through spirituality" is the basis of Man's adaptability, i.e., of his existence.

Therefore, the religious Man of the Middle Ages knows little, but better understands the world and his place in it, while *a contemporary Man* knows a lot but does not understand his place in the world and the purpose and meaning of his existence, that is, under the lack of spirituality he becomes more and more deadaptive and develops following the positivistic paths rather than the canons of Life evolution. This is a deeply erroneous way and is the basis of all anomalies and negatives of *being* of a *contemporary Man*.

By losing the religious spirituality and accepting a positivistic view on the world, a Man substitutes his spiritual way of adaptation by positivistic, in essence, and a "negative" way of adaptation which is not adequate to his nature, that is by the adaptation directed towards Man's domination over nature. In so doing *a contemporary Man*, mythologizing his creative constructive activity, illusively "deifies" himself. In reality, he turns into a materialized Man, with one-dimensional and controlled by external factors consciousness. In other words, by losing his spirituality, that is deadening his sense of integrity, by estranging from spirituality and its values, he strives to create a certain positivistic world which functions following the canons of logic and rational knowledge. The idea of Man's domination over nature and in social life, which gives birth to spiteful aggressiveness and insatiable money-grubbing, leads to his estrangement from the external world. This is a fundamental negative quality of positivistic Man, which is the essence of crisis of a *contemporary Man*, since his *being* goes beyond the scope of the canons of Life evolution and becomes catastrophic.

To say more, the loss of spirituality, that is degradation of Man's ability to understand the universe, leads inevitably to morbidness of psyche of a *contemporary Man*. The loss of the sense of integrity, directly connected with spirituality and being a fundamental human sense, causes his mental disorder and Man's actions become mindless.

In reality *a contemporary Man*, not understanding the problematic character of his further existence in the world, tries to solve the question of the lack of spirituality in a roundabout way, namely through reformation of the sphere of his *being*, i.e. using different social, ecologic, ideological and scientific-technical myths. But that is the way which leads to nowhere. A contradictory way of Man's development through the pattern:

$$\textit{Homo sapience} \longrightarrow \textit{[...]} \longrightarrow \textit{nihilist} \longrightarrow \textit{positivist}$$

is a dead-end way of mankind's development, since positivist is characterized by the loss of the sense of world integrity, mercantile attitude towards the external world and civilized (or technological) barbarism. As was noticed above, the atomic, hydrogen, chemical, biologic weapons created by *a contemporary Man,* as well as militarization of space can serve as a vivid example. So, we have a good cause to say that *a contemporary Man*, losing his reason, already became a mortal disease of Life. Under the lack of spirituality, his evolutional potency being exhausted, he has found himself in the deadlock of evolution.

A Man does not sense and does not know the image of the future, the purpose and meaning of his existence in the world. Losing his adaptability, he lives in conditions of ambiguity and confusion that is he internally senses the problematic character of his further existence and feels fear of the future. That is particularly true of the eternal question; what is death and what awaits a Man after his death. In other words fear of the future (or evolutional fear) ever increasingly dominates over him, rather than the archetypical sense of the integrity of existence, being in normal state a result of the archetypical phenomenon of Self.

A psychic crisis which arose because of the lack of spirituality and is expressed in mental splitting, as well as decadence of *a contemporary Man* have opened extremely broad opportunities in controlling his behavior which, in particular, gave birth to the illusion of new Utopian societies, such as the communist society, society of material abundance and prosperity, society of sustainable development, elite society, society of people's self-management, society of risk, etc. Besides, in conditions of search for new spirituality, such things like anti-scientific Gnosticism-theosophy, astrology, theories of paranormal phenomena, etc are revived more and more.

It should be noted that the psyche of *a contemporary Man* in its nature is archetypical, and in normal state, that is when a Man lives as a spiritual creature, it is always integral, integrated and is perceived as a soul. Only in the process of intellectual evolution as a result of spiritual degradation, caused by continuously inhibiting influence of knowledge, especially technical and technological intellect, he (Man) had lost this fundamental quality, and his psyche suffered a deepening split into consciousness and unconscious. This crack, this deep split of psyche manifested itself as far back as in the time of Renaissance in the conflict between knowledge and faith.

In the meantime civilization further and further estranged a Man from his archetypical basis of existence. And the conflict between knowledge and faith, responsible for psyche split, becoming more escalated, destroys the psychological and psychosomatic integrity of a Man. In so doing, the historical fact of secularization, that is the liberation of various spheres of society, individual and social consciousness from religion, may be taken as the beginning of the process of splitting. Thus, we find ourselves in front of the abyss which appeared between nature and mind, between the sphere of unconscious and consciousness. In this connection it is important to note that interrelation between religious spirituality and knowledge is of syncretic character. And thus, a loss of each of them turns a Man into a monster (ascetic, fanatic, narcissus, one-dimensional Man, Man-robot, inform–Man, econom–Man, etc.).

Thus the spiritual crisis of *a contemporary Man*, accompanied by an unordinary state of consciousness, psychic splitting, principally new kind of fear and phenomena of estrangement, moral degradation, economical, ecological, social and technological catastrophes, in one word, by his decadence, is a real eschatological problem of the present. It is directly connected with the crisis of Self, the archetype of integrity, as a result of which, the way of life of *a contemporary Man* is more and more mystified, that is due to the loss of sense of the integrity of the world his self-insight and comprehension of the world become more erroneous. As a result a Man often misunderstands and misinterprets his life situation and loses adaptability. Thus, *a contemporary Man* is evolutionally ill and his diagnosis: spiritual crisis, i.e., the loss of the highest "Ego", composing the spiritual essence of a Man; positivistic view on the world; psychic splitting (dual personality); mental disorder (loss of reason); moral bankruptcy; intensification of social conflicts; deadaptation; fear of the future.

Thus, *a contemporary Man* degenerates into a positivist, mental disorder and nihilism, being characteristic of the latter.

The algorithm of negative aspects of spiritual crisis can be presented as follows:

Spiritual crisis (loss of spirituality)   psychic disorder (psychic splitting) neurosis   estrangement; negative adaptation   stress; functional disorder of organs and organism as a whole   fear, including fear of the future   "epidemic" of transformation of a contemporary Man into positivist   increase of risk factor in the existence of mankind   decadence of a contemporary Man apocalypses.

But in contrast to ordinary psychopathological diseases connected with biologic, traumatic causes, the spiritual crisis also has a positive, healing effect. That is why the true psychosis of biologic (organic) origin and negative aspects of spiritual crisis should be identified separately. And that is a rather complex problem which requires a qualitatively new level of understanding of Man's psyche, since their symptoms are almost identical and indistinguishable. Until now, this problem is often not accounted for in medical practice and remains open. According to STLife, the problem is solved rather simply, since the psychosis of organic nature is directly connected with individual unconscious, whereas the spiritual crisis – with collective unconscious.

The algorithm of positive aspects of spiritual crisis (lack of spirituality) may be presented as follow:

Spiritual crisis (lack of spirituality)   psychic disorder (psychic splitting) neurosis   estrangement; negative adaptation   stress   fear, including fear of the future (or the evolutional fear of a contemporary Man)   actualization of creative, constructive activity of people directed towards the spiritual renewal which is a powerful healing remedy and bioenergetic source of further evolutional development   new views on the world and new values and aspirations   mental enlightenment, self-sufficient Man.

The real appearance of the positive or negative aspects of spiritual crisis is determined by the way of life and level of intellectual development of the individual, including his mystic gift. The most ancient and vivid example of use of positive aspects of spiritual crisis is shamanism which was preserved until present day.

Thus one can conclude from above stated that *a contemporary Man* in his evolutional development has entered the zone of renovation, i.e., he is being transformed into a positivist whose *being* is nihilistic, mercantile and destructive, since his psyche even in early childhood absorbs positivism, and this process progresses. Figuratively speaking, the spiritual atmosphere is more and more being saturated by positivism, individualism and nihilism, which together threaten the existence of Life. That is why, to preserve Life on the Earth a Man has to become absolutely different, that is: he has to regenerate into a new species – a self-sufficient Man, whose world outlook, *being*, way of Life will be directed not to positivism, but to synergetic world outlook, scientific religion and aspiration for archetypical ideal – absolute perfection, the Absolute. In other words, the Life evolution currently is leading mankind towards bifurcation into a positivist and a self-sufficient Man. This radical change of the system of values and life orientation of a Man is the evolutional destiny of mankind.

About us

Kemal Atayev, born 1991, is currently a student of the University of Nottingham. He has graduated from the University of Birmingham with an undergraduate degree in Accounting and Finance and is currently finishing his postgraduate degree in Risk Management at Nottingham.

Ovez Atayev, born 1994, is a student of the University of Birmingham. He is currently doing his undergraduate degree in Physics and is planning to pursue an academic career in Physics afterwards.

About The Author

ORAZGELDY OVEZGELDYEV (pen name ORAZ TURKMEN)
Professor, Academician, Doctor of Physical and Mathematical Sciences
(1936 – 2008)

He was born in 1936 in Turkmenistan (USSR).

In 1958 he graduated with honours from the Physics department of Physics and Mathematics Faculty of the State University of Turkmenistan.

Later in 1958 he was assigned to work at the Academy of Sciences of the Turkmen SSR and worked there until the end of his life - for 50 years.

Orazgeldy Ovezgeldyev began his career in the laboratory of ionospheric research of Physico-Technical Institute of the Academy of Sciences of the Turkmen SSR and has gone all the way to become the President of the Academy of Sciences of the Turkmen SSR.

In 1962 he defended a thesis for the degree of the Candidate of Physical and Mathematical Sciences in the subject of Radiophysics at the Tomsk State University, Russia.

In 1971 he defended a thesis for the degree of the Doctor of Physical and Mathematical Sciences in the subject of Radiophysics at Irkutsk State University, Russia.

Since 1978 he was a full member (Academician) of the Academy of Sciences of the Turkmen SSR.

Orazgeldy Ovezgeldyev was the chief editor of the journal "Izvestiya of the Academy of Sciences of the Turkmen SSR: the series of "physico-technical, chemical and geological sciences".

He is the author of above 300 scientific works in radio physics, geophysics, solar-terrestrial physics, astrophysics and science of science.

Orazgeldy Ovezgeldyev made a significant contribution to the development of new promising scientific directions in the Academy of Sciences of the USSR and is a renowned scientist in the fields of:

1) Solar-terrestrial physics;
2) Ionospheric physics;
3) Near-earth space environment;
4) Science of science.

In recent years the range of his scientific interests became much wider: synergetics; physical problems of ecology; economic, social and environmental aspects of scientific and technological progress. Two new scientific directions are successfully developing because of the scientific research of O. Ovezgeldyev:

1) Physics and structure of turbopause;
2) Spectrophotometry and Aeronomy of meteors;

In the 1960s he discovered the synergetic structure in the turbopause of the Earth's atmosphere, which became the beginning of his work to create a Synergetic Theory of Life.

The monograph "Synergetic Theory of Life" (book of Life's renewal) was first published in 2001 in Ukraine. The second edition of the monograph in Russian was published in 2009 in Kiev by the Publishing House "Naukova Dumka" of the National Academy of Sciences of Ukraine.